# The Economics of
# Tropical Agriculture

# The Economics of Tropical Agriculture

## Dr. BERND ANDREAE

Professor of Agricultural Production Economics,
Technical University Berlin

with 33 figures, 48 tables

English Edition Edited by   Jean Kestner, M.A.,
Commonwealth Bureau of Agricultural Economics
Oxford.

COMMONWEALTH AGRICULTURAL BUREAUX

Commonwealth Agricultural Bureaux
Farnham Royal
Slough SL2 3BN
England
Tel. Farnham Common 2281
Telex 847964


ISBN 0 85198 453 3

"The authorized translation of Bernd Andreae, Landwirtschaftliche Betriebsformen in den Tropen, 1st edition, © 1972, Verlagsbuch-handlung Paul Parey, Hamburg and Berlin"


Printed in Great Britain by Robert MacLehose and Co. Ltd.,
Printers to the University of Glasgow

# Foreword

In the last few decades German agricultural science has been faced with the new task of dealing with research and teaching relating to agricultural problems in developing countries. The problems with which they are dealing have become decisively important in recent years and their significance is not merely a question of economics.

Even so development policy is very much an economic policy and developing countries are agricultural countries. Development policy of such countries is, thus, to a considerable extent agricultural policy and must draw on knowledge of the production elasticities of different elements of agriculture, the farm and the plantation, to find the right policy measures and priorities. From this knowledge it must select and implement economic instruments which can raise productivity and income and which are appropriate in their nature and extent to a particular place and a particular period of time.

However, one farm is not exactly like another and each plantation is also different and a bewildering number of types of agricultural enterprises arise because of ecological and economic differences. To gain a systematic picture of these manifold forms it is necessary to understand the farm management relationships and to deal separately with factors not related to management. Further, the morphology and physiology of the most important types of enterprise must be investigated and their problems and possibilities of development have to be recognised. It is only on the basis of such insights that the main objectives of development policy can be achieved through properly directed measures and that their effects can be estimated more realistically than in the past.

For this reason the concern and content of this book are the system, morphology, physiology, location and evolution of the most important types of farming in the tropics. Global and regional statistics were not very informative for this purpose and farm level statistics were rarely available. The quite extensive literature, particularly in English, French and Dutch, mostly keeps to a fairly descriptive level and does not place much emphasis on functional or causal relationships which are of particular importance for my purpose. My interpretations are therefore primarily based much more on what I have seen for myself, on geographical comparisons and on comparisons of economies at different stages of development. In the course of my studies I have travelled not only to developing countries, such as Egypt, Ethiopia, Kenya, Mozambique, Nigeria, Rhodesia and Tanzania but also to countries at higher stages of industrialization like the Republic of South Africa and the USA. The understanding of types of farming under similar ecological but differing economic conditions is one of the most important keys to an understanding of the obstacles to development and of development problems, possibilities and incentives. It provides the answer to the question "stagnation, evolution or revolution—which is to be the type of farming in the tropics?"

My wife has made a small direct contribution to this book but a very large indirect contribution.

Berlin-Dahlem, Winter 1971/72                                              Bernd Andreae

# Preface to the English Edition

The preparation of an English edition of this book six years after the original has given me the opportunity to include additional material based on further visits to tropical countries, studies of recent literature and consultancy and seminar activities. I am therefore grateful to my German and English publishers for giving me the opportunity to fully revise and extend this book and to publish it in a world language in which it can reach a wider circle of readers and I would like in particular to thank Mrs. Kestner for assistance in editing and translating the English edition.

Naturally the author of a book likes to refer back to his own shorter earlier studies. I would firstly like to thank the Institute for Scientific Co-operation in Tübingen for making available my papers in their documentation text "Economics". In addition I would like to thank Dr. W. Junk bv. Publishers, The Hague, for allowing me to reproduce some parts of my contribution to Vol. 13 of their "Handbook of vegetation science". Finally Dr. Wolf Tietze, the editor in chief of the "Geo-Journal, Wiesbaden" generously gave me a free hand to make use of both my articles in issues Nos 2 (3) and 2 (4).

I hope then, that this book will find its rightful place among development theory studies and will help agricultural countries with their development problems. Much of the tropics falls into this category and this is a matter of life and death for whole populations.

Berlin-Dahlem, August 1978                                    Bernd Andreae

# Contents

# List of Tables

# List of Figures

# Chapter 1

# Introduction

## Between Famine and Food Surpluses

### Idealistic hopes and hard realities

Both developed and developing countries face agricultural problems. The European Economic Community's agricultural problems led to the Mansholt plan for reducing the farming area which met with a mixed reception. The agricultural problems of developing countries have led to trends in malnutrition and famine which give statisticians fears for the near future. Non-experts find it difficult to understand how, in a world which has grown smaller as communications have improved, a permanent and dangerous food shortage can co-exist with food surpluses. Experts, however, know the obstacles to achieving a balance. The Greek Professor of agricultural economics Euthymios Papageorgiou (1969), suggested a new solution to agricultural problems both in industrial and developing countries. His basic ideas merit consideration because of their positive aspects although their extreme originality perhaps makes them fall somewhere between enlightened insight and impracticable utopianism.

The root of the agricultural dilemma of many industrialized countries lies in the fact that their food production increases faster than their food consumption. In the EEC, this imbalance creates huge surpluses of soft wheat, butter, cheese etc. These surpluses exert a pressure on the market and become a burden for the state budget, the producers and the consumers. Thus, the prices of agricultural products are high compared with international prices but low compared with production costs. Costs are high because of outmoded production methods and agricultural structure. The result is inadequate agricultural incomes. Mansholt (1968) proposed that the EEC should try to restore the parity of farm and non-farm incomes for the period up to 1980 by transferring 5 million farmers out of agriculture, letting 5 million ha of farmland fall out of cultivation and by slaughtering 5 million high-yielding dairy cows. In addition to these measures small family farms should be replaced by large farm businesses. Thus producers could expand their land area, their livestock numbers and their machinery and so depend less on intensive types of production for adequate income.

Those working in agriculture must increase the physical volume of their output if parity is to be achieved between agricultural incomes and those of industrial workers. Mansholt saw the solution of the EEC agricultural problem in the reduction of overall production, a smaller cultivated area, fewer dairy cows and less intensive farming. The West German government's agricultural programme has followed the principles of the Mansholt Plan but less rigidly. It sets a limit to its extent and duration and is evolutionary in character rather than revolutionary like the Mansholt Plan.

Looking at the developing countries, Papageorgiou dismissed such a reduction of production as noneconomic, unjustifiable and antisocial. He stressed the fact that 25 million people die every year of hunger and that according to FAO studies two thirds of the food produced is consumed by one third of the world's population while the remaining two thirds are undernourished. World population at the time of Christ's birth amounted to 250 million, but following the biblical saying "Be fruitful, multiply, replenish the earth, and subdue it", it now stands at 4.04 billion, and will probably reach 6 billion people by the year 2000. According to FAO estimates, by the year 2050 the world population could reach 14–15 billion and, food requirements would thus be six to eight times greater than those for 1960.

These huge numbers demand an effort from the whole world if major disaster is to be avoided and there must be no reduction in the agricultural production of EEC and North America. Papageorgiou considers the only way out of this dangerous situation to be the establishment of a World Fund against Starvation under the auspices of UNO or FAO. The 1969 FAO annual report showed that surplus cereals in USA reached 22.1 million tons, in Canada 22.6 million tons, and in EEC 10.0 million tons. Stockpiled butter in USA in 1969 reached 55 000 tons, in West Germany 104 000 tons, in France 151 000 tons, in Holland 40 000 tons, and in New Zealand 52 000 tons. Even if all stocks necessary to cover fluctuations of production are allowed for, huge net surpluses still remain.

*Fund against starvation*

The Fund against Starvation should collect all these food surpluses from industrial countries and distribute them among the hungry people in the poor countries. In this way the stockpiled food of industrialized countries would cease to exert pressure on the market and burden the state budget, producers and consumers. Instead it should be put at the disposal of the 'Fund against Starvation' which could either buy it or accept it as a donation to food aid from the country offering it. There are a number of ways in which such a fund could be financed. Possibilities include a percentage of the general budget of UNO member-countries or of their special military budgets. Other alternatives are contributions from large and prosperous countries such as the USA, Great Britain, Canada, the Soviet Union, Australia, Germany, France, etc. or special contributions from international organizations and institutions such as the International Bank, International Monetary Fund, FAO, OECD, EEC, EFTA, COMECON, Ford Foundation, Rockefeller Foundation etc. and contributions from philanthropists and charitable organizations.

The Fund executive board should be composed of delegates from the donor countries and organizations. The success of such a Fund against Starvation would depend on matching the demand of undernourished peoples for food with the agricultural surpluses of industrialized countries, on having the cash available to buy up available food surpluses and on the finding of necessary funds to back the venture. Taking into account the development of population, agricultural production and prosperity in hungry agricultural countries Papageorgiou estimates the monetary needs of the Fund at $27.2 billion per year (the Federal German budget for 1969 was $22.8 billion). He believes that such a sum could be raised, and points out in support of this view that the USA, EEC and other countries are currently spending more than $10 billion per year for foreign aid and more than $500 billion has so far been spent on space exploration.

The practical success of such a programme would not only provide outstanding assistance to hungry underdeveloped countries but would also be highly important for EEC agriculture. The Community could not only keep food production at today's levels and allow surpluses to be stockpiled by the Fund against Starvation, but a further increase in production would be possible and urgent in the fight against hunger. Mansholt's demand for immediate reduction in production could be reversed. With a need to increase production the current EEC agricultural structures which are so severely criticized as out of date could once more become acceptable and it would no longer be necessary to expect 10 million farmers to change their way of thinking. Instead of making violent changes which could have dangerous effects, methods which have worked successfully in the past could continue to be used. The greater the production required, the larger the cultivated area and the number of farms and farmers could be.

However, as with every great goal, the long road to the Fund against Starvation is strewn with obstacles. Even if all the budgetary needs were met, success would not be guaranteed. Papageorgiou should, perhaps, have asked Onnasis whether the merchant fleet of the whole world would be sufficient to transport the cereal surpluses from North America to the poor countries of Africa, Latin America and S.E. Asia. The cost of transporting as much as possible through the international seaways to large ports like San Salvador, Dakar, Kinshasa, Bombay or Rangoon would not be prohibitive. There are, however, larger obstacles and difficulties. Port facilities in tropical countries are not always adequate for large scale handling of cereals. In many cases loads have to be stored and are vulnerable to rats, insect pests and rot. These dangers cause much more damage in tropical countries than in temperate ones, partly because of lack of facilities, but mainly because of climate. Thus in Nicaragua and Haiti about 30% and 47% of all maize crops are destroyed by parasites and storage conditions.

There are just as many problems in transporting cereals to the interior of under-developed countries as the transport network is nowhere equal to that in developed countries. There are insufficient roads, railway lines, or waterways. About 200 million people still live on one quarter of the earth's total surface in almost the same way their ancestors lived in the neolithic period. This is especially so in inland tropical regions. They wander from place to place, they practise exhaustive cultivation, they ignore civilization, they are illiterate, they know nothing about crop rotation, fertilizers, work-animals, tractors, sails or wheels. Away from the few roads and railway lines all goods are transported on people's heads.

It is difficult to see how cereals from the Fund against Starvation could be carried along hundreds of miles of narrow roads through to the last remote village or mountain hut or how one could guarantee that Timbuktu and Dakar would get the same quantity based on FAO's estimates of food requirements per head. How could one make sure that those who live far from the seaways in the interior of a country would receive the same store of supplies as those living in easy reach of the Congo waterways or that those in very mountainous areas received as much as those living in the plains? Cereals may seem easy to transport but those who know the interior of tropical countries know this is not always the case.

The transport of powdered milk could be a more economic proposition and possibly the Fund against Starvation could help in two ways by collecting, transporting, and distributing this type of food. Industrial countries face the problem of overproduction of milk while underdeveloped countries are short of albumen. An effort would have been made to solve both these problems much sooner if the production of powdered milk were not so expensive. There would, however, also be problems in sending powdered milk from Europe to countries in humid tropical zones. All the year round the average relative humidity of the air in Axim, on the

Ghanian Coast at 6 o'clock in the morning is above 89%. Although it is costly and difficult to extract water from milk in Europe, it is easy for that powdered milk to absorb humidity in Ecuador. Today's cans and packaging materials are not fully airtight as is necessary for transport and the first stages of distribution. Storage and preservation conditions in a hut in Africa or S.E. Asia are unbelievably bad. Even if food aid from the Fund were only transported as far as urban centres, the costs of distribution, storage and preservation of food would be prohibitive.

## Production is easier than distribution

Those are some thoughts, perhaps not the most important, on Papageorgiou's plan. They do, however, raise a very important factor which the Greek overlooked—the incomplete communications network and administrative infrastructure of under-developed countries. If the world food problem were only a matter of production it could perhaps be solved by a World Fund against Starvation. It is, however, not so much a problem of production as one of distribution. There is no way round this bitter and undeniable truth.

## Bibliography

Andreae, B. (1972) Between hunger and surpluses. Ideal prospects and hard realities. *Hellenic Agricultural Economic Review* 8 (2), 157–162.

Andreae, B. (1974) [Which farm will survive? Preserving the farm business by developing it.] Welcher Hof wird überleben? Betriebserhaltung durch Betriebsentwicklung. Hamburg, German Federal Republic; Paul Parey, 184pp.

Andreae, B. (1977) [Agricultural geography.] Agrargeographie. Berlin; New York, USA; Walter de Gruyter 332pp. [English language edition forthcoming, New York 1980.]

Andreae, B. (1978) [Agricultural regions under local stress.] Agrarregionen unter Standortstress. Kiel, German Federal Republic; Verlag Ferdinand Hirt, 78pp. [Geocolleg. No.6].

Blanckenburg, P. von (1970) [Protein supply as the core of the world food problem.] Die Eiweissversorgung als Kern des Welternährungsproblems. *Zeitschrift für Ausländische Landwirtschaft* 9 (1), 1–23.

Boguslawski, E. von (1969) [The technology of crop production and meeting human needs.] Die Techniken der Pflanzenproduktion und die Deckung des menschlichen Bedarfs. Rome, Italy; IIIème Congrés Mondial de la Recherche Agronomique, 1st–5th December, 1969. 23–69.

Mansholt, S. (1968) [Report on the Common Agricultural Policy, Parts A-E.] Dossier concernant la politique agricole commune. Partie A-E. [*Publication*] *European Communities* No.COM (68) 1000 Parts A–E.

Papageorgiou, E. (1969) The present currents of European agriculture and the world food problem, *Hellenic Agricultural Economic Review* 5 (2).

Ruthenberg, H. (1970) [The world food problem.] Das Welternährungsproblem. *Freiheit und Ordnung* No. 68, 34pp.

Teo Chris, K.H.; Atanasiu, N. (1975) Increased land productivity or expanded area as a means for increasing food production in developing countries. Workshop on Energy, Resources and the Environment. Penang, Malaysia.

Witt, M. [Repercussions of the world food situation on food production in Western Europe.] Rückwirkungen der Welternährungslage auf die Nahrungsproduktion im westlichen Europa. *Jahresheft der Albrecht-Thaer-Gesellschaft* No. 14, 96pp.

# Chapter 2

# Minimum Cost Combination in Agriculture with Special Reference to Developing Countries in Tropical Areas

Theodor Brinkmann (1922), one of the great German farm management theorists, once divided the multitude of individual farm management problems facing the farmer into three basic groups: (1) those concerned with the selection of the most economical equipment and production materials; (2) those concerned with the use of production equipment and materials so as to obtain the highest possible yields; and (3) those concerned with achieving maximum utilization of products.

This chapter deals with the first of these three groups of questions. It starts by looking separately at the three classical production factors land, labour and capital and goes on to distinguish between various forms of capital goods.

## Theoretical Principles

### The law of diminishing marginal returns

The minimum cost combination is related to the law of diminishing marginal returns which states that: ". . . the increase in gross profit in agriculture does not correspond to the increase in operating expenses, i.e. the increase in gross profit per additional unit of value invested after a certain point tends to diminish continually until it finally disappears. For a private enterprise, where the gross profit and the operating expenses are represented by monetary values, this means that the difference between the monetary value of the units of input and the monetary value of the corresponding units of output will also gradually decrease and will become negative before the potential limit in yield increase has been reached. For this reason it can never be profitable to strive for maximum yields, let alone to increase intensity of production indefinitely. In private enterprise the maximum permissible increase in operating costs is reached when the value of marginal input is covered by marginal yield, i.e. when the last unit of value invested is covered by the last unit of profit achieved . . ." (Brinkmann 1922 p.32–33).

This law of diminishing marginal returns becomes obvious wherever an increase in input involves only one production factor, while the input of the other factors remains the same, and wherever an input is increased disproportionately. Figure 1 plots marginal costs per unit of fertilizers against the marginal value of the potato output. The curve showing the marginal returns is roughly hyperbolic, descending from the top lefthand corner to the bottom righthand corner at first rapidly and then more gradually. The marginal input is a straight line parallel to the abscissa because the farmer pays the same sum of money for each fertilizer unit, regardless of how many units he applies. The figure shows how the law of diminishing marginal returns makes it possible to determine the optimum degree of intensity and uses the application of fertilizers to potato crops as the first example. The dotted line and curve in Fig. 1 show the relationship between input and output in the USA. The last fertilizer unit per ha as shown by the straight line A1 is to be regarded as the marginal input. One fertilizer unit therefore costs US $23.72. The marginal output curve is shown by the hyperbola E1. With the conditions shown in this case the highest permissible intensity in the application of fertilizer is five fertilizer units, since at this point (P1) the marginal input and the marginal output are balanced. If the application of nitrogen fertilizers is increased beyond the point P1, financial losses will occur, since the marginal input is then no longer covered by a corresponding marginal output. If, on the other hand, less than the optimum units of fertilizer are applied, e.g. for two units per ha less, the profit potential is not fully exhausted. The net return, the important element for the farmer, is shown by the area integral between the line A1 and the curve E1. The fact that this area integral is smaller if only three fertilizer units per ha are applied than it is if five fertilizer units per ha are used shows that the net return obtained is smaller if less than the optimum amount of fertilizer is used.

Fig. 1
The law of diminishing returns, illustrated by output of potatoes per unit of fertilizer. (Source: Heady 1957, p.36 and p.100)

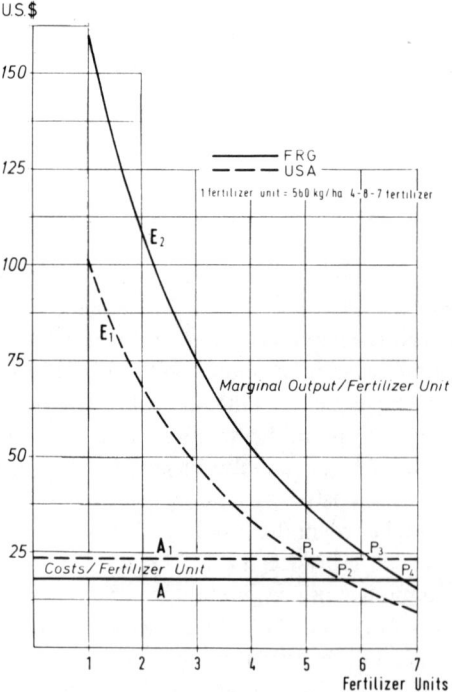

The graph also shows how the intensity of fertilizer application must vary if the price-cost relationship changes. Let us assume that the cost of mineral fertilizer sinks to the level shown by the line A2. The point of intersection of the marginal input line and the marginal output curve now lies at P2. This means that about 5.8 fertilizer units per hectare must now be applied for optimum results. Or let us suppose that though the cost of mineral fertilizers remains steady at the level of A1 the price of potatoes rises to the level shown by the marginal yield curve E2. This again makes the intensified use of fertilizers possible. About 6.2 fertilizer units per ha should now be applied. This is shown by the fact that the point of intersection of the marginal input line and the marginal output curve now lies at P3. The greatest increase in the number of fertilizer units applied per ha is necessary when the prices of agricultural products increase and those of mineral fertilizers simultaneously decrease. In this case the input-output ratio is illustrated by A2/E2 (American rates for growth of crop yields and 1963-64 West German prices and costs). The point of intersection is now at P4 and optimum fertilizer application would therefore be 6.7 units per ha. Fertilizer application should therefore be more intense in the Federal Republic of Germany than in the USA. We shall be going into this in more detail later.

Fig. 2
The law of diminishing returns: output of sugar beet per unit of labour for hand hoeing. (Sources: Prjanischnikow 1930, p.145; Asmis 1919, p.457 ff; Hawmann 1911, p.661 ff; Marcard 1890, p.155)

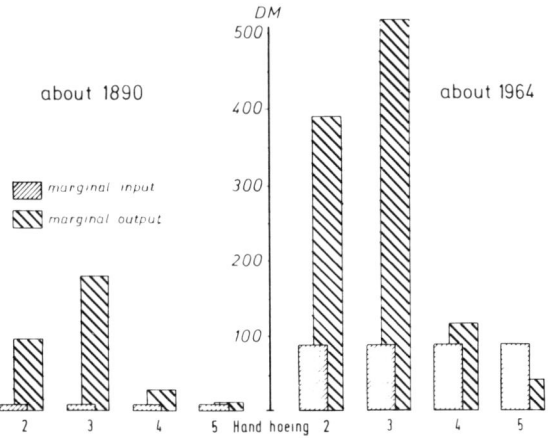

The law of diminishing marginal returns also applies to the input of labour. For instance, it is not possible to increase the input of labour for a given area of sugar beet indefinitely and to achieve a correspondingly steady increase in yields. A certain minimum of tillage and cultivation is necessary if anything is to be harvested at all (minimum labour intensity). An increase in the input of labour makes for an increase in yield, but after a certain limit has been reached the increase in marginal yield gradually falls off until it disappears altogether. The first hoeing usually leads to a greater increase in yield than does the second. And the increase in yield caused by the second hoeing is greater than that brought about by the third. If fourth, fifth and sixth hoeings are undertaken the corresponding increases in yield will be progressively less. For reasons of profit the number of hoeings carried out should of course, be such that the cost of the last hoeing is just covered by the additional yield it produces, i.e. marginal output should equal marginal input. Figure 2, which is based on experiments in the Kiev area, shows that in 1890 five hoeings were appropriate for sugar beet whereas in 1964 only four were worth while. The figure uses Russian rates of increase in output and the relevant German price-cost conditions.

**Labour productivity versus land productivity**

The comparative costs of the three factors of production determine the relative proportions in which they are used. However, as these comparative costs change considerably in the course of national economic development, and as there are great differences from one country to another, the minimum cost combination also varies greatly. Thus the proportions of land, labour and capital which combine to produce the best results also vary considerably. The degree of intensity of cultivation selected depends on whether it is considered more important to make optimum use of land to obtain high land productivity, or whether it is more important to make the maximum use of labour and capital to obtain high labour and capital productivity.

In this connection F. Aereboe says (1923 p.291 ff): "The relationship between the importance of utilizing labour and capital on the one hand and that of utilizing land on the other is not the same at all levels of agricultural development. Where land is expensive the question of its utilization is more important than where it is cheap. Where the price of land is high in comparison to that of labour and capital the utilization of the land will have to be given priority over the utilization of labour and capital. If, on the other hand, labour and capital are comparatively expensive, then farm management must be oriented chiefly towards the optimum utilization of labour and capital.

"At the lowest stage of agricultural development uncultivated land as such has no price and thus also no value. Uncultivated virgin land is here available in abundance. Not until land reclamation work or the advantages of road, railway and other construction have become inseparably connected with the land does it come to have a price and thus also a value. At first the price is low, but it rises during the course of development and the question of land utilization grows correspondingly in importance. At the lowest level of agricultural development the farmer must therefore concentrate exclusively on utilizing the labour and capital invested in the land to the highest possible degree. Even where the land has already gained a certain low value the paramount objective is still to achieve the highest possible yield in order to cover the cost of labour and capital invested. The interest in achieving a high yield per hectare of land still remains entirely in the background . . .

"At the lowest level of agricultural development the amount of land cultivated must suffice to ensure that the labour and capital invested alone provide as high a crop yield as possible. The amount of virgin land required for this is entirely immaterial so long as this aim is achieved.

"Indeed, up to a certain limit the crop yield achieved increases with the area of land in which a certain amount of labour and a certain amount of capital are invested. This means that extensive agriculture leads to high labour and capital productivity. The following thesis is thus of the greatest importance for us: 'In agriculture labour and capital considered on their own are, up to a certain limit, more productive the more extensive the cultivation undertaken.'

"What, on the other hand, is the situation with regard to the yield achieved per unit of area when investment varies? How does land productivity stand with regard to varying degrees of cultivation intensity? It is obvious that the situation is reversed here. The higher the input of labour and capital per unit of area, the higher, up to a certain limit, is the yield of this area. The better the tilling, fertilizing, sowing and reaping, the greater, up to a certain limit, and assuming all other conditions to be equal, is the crop yield that can be achieved per unit of area.

"Thus optimum land productivity requires intensive cultivation, while optimum labour and capital productivity require extensive cultivation. Since land as well as labour and capital are necessary for agriculture, a compromise must be found between the two opposing interests. The form this adjustment will take must depend on the relation between the price of land on the one hand and the cost of labour and capital on the other. The more expensive land becomes in relation to labour and capital the more intensive cultivation must become. The more expensive labour and capital become in relation to land, the more extensive cultivation must become."

**The marginal productivity principle**

The basic conclusions to be drawn concerning the minimum cost combination in agriculture are therefore extremely simple to summarize. The farmer must strive to combine his operating resources in such a way that each production process can be carried out at the least possible cost. For this purpose the three factors of production, land, labour and capital, must be combined in a specific proportion—the minimum cost combination. The proportions will depend on the cost of each of these three production factors, namely ground rent, wages and interest.

The reason for this is that the input of each factor of production is subject to the law of diminishing marginal returns. Thus, if the input of capital is increased while that of land and labour remains the same, then the additional returns from each additional increase in capital will grow progressively smaller. The input of capital must not be increased beyond the point at which the cost of the last unit of capital invested is just balanced by the corresponding increase in the returns from extra production. The same principle applies if the input of land is increased while the input of labour and capital remains the same or if the input of labour is increased while the input of land and capital remains the same. It will always be observed that the greater the input of one factor of production in relation to the other two, the greater the extent to which its marginal productivity decreases. The farmer must always try to increase the input of each factor of production just sufficiently to ensure that its marginal productivity covers its costs. The minimum cost combination has been achieved when the marginal returns on all three production factors are proportionally equal to their marginal costs.

The consequence of this is that, other conditions being equal, expensive factors of production must be applied sparingly. On the other hand, the largest possible quantity of the cheapest factor of production should be used in the production process, since as its cost is low it is still economic to use it when its marginal productivity is low. Thus the input of a cheap factor can be greatly increased in spite of the law of diminishing returns.

## The Minimum Cost Combination at Different Stages of Economic Development

It is thus possible to deduce the optimum combination of factors of production at different stages of national economic development directly from the theory of marginal productivity. This can be demonstrated by a table worked out by H.-H. Herlemann (1954) which is reproduced below. This chapter draws on his basic ideas and quotes from them.

Table 1   Changes in the combination of agricultural production factors in the course of economic development.

| Stage | Description | Land | Labour | Capital | Order of development |
|-------|-------------|------|--------|---------|----------------------|
| 1 | Thinly populated agricultural countries | + | + | − | |
| 2 | Beginning of industrialization | + | − | − | New World |
| 3 | Industrialized agricultural countries | + | − | + | |
| 4 | Industrialized countries | − | − | + | |
| 5 | Agro-industrial countries | − | + | + | Old World |
| 6 | Overpopulated agricultural countries | − | + | − | |

Source: H. H. Herlemann, 1954, p. 335 f.

The table shows the relative shortage of a factor by a minus sign and its abundance in relation to the other two by a plus sign. Stages 1 and 6 represent agricultural countries, i.e. developing countries in which no significant differentiation in the economy has so far taken place. Stages 2 to 5 illustrate the result of various levels of industrialization which are usually characteristic of economic development. This process of industrialization can take two fundamentally different courses, depending on whether it takes place in a thinly populated country or in a country with a high population density.

**Development of a thinly populated agricultural country into an industrialized country**

Let us first consider changes in minimum cost combination arising during economic development in thinly populated agricultural countries with fewer than 60 inhabitants per 100 ha of farmland.

*Thinly populated agricultural countries*

This group of countries includes most of the Central and South American countries and many African countries, such as Ethiopia, Sudan and Zaire. As land is abundantly available in these countries it is also cheap. Manpower is also cheap at least in the early stages of development, but all capital goods purchased from industry are exceedingly expensive. It is therefore possible to do without high land productivity and without high labour productivity, but it is very important to achieve high capital productivity. The result is that land is used extensively and that economies are achieved in the use of capital goods wherever it is possible to achieve the desired effect by an increased input of labour. The agricultural system as a whole must be regarded as extensive.

*Initial stages of industrialization*

The next stage is marked by the beginning of industrialization and Canada and Australia may be taken as typical examples. Land is still abundantly available but the labour available to agriculture is being reduced as a result of competition with industry. Capital goods are less scarce. The relation between the agricultural production factors is now altered by the fact that land and labour are growing more expensive while capital is growing cheaper.

Labour productivity must now be increased above the level in Stage 1, while it is possible to relax the emphasis on capital productivity. The input of labour is therefore reduced by increasing the input of capital. The agricultural system as a whole must still be regarded as extensive.

*Industrialized agricultural countries*

In the third stage of development, for which the USA may be taken as an example, land is somewhat scarcer but still relatively cheap. As a result of the industrialization process the cost of labour has greatly increased, while the cost of capital goods and credit has decreased. As a result labour productivity must again be greatly increased, while capital productivity may be allowed to relax even further, i.e. agriculture must be labour extensive and capital intensive.

To quote Herlemann (1954) again, the changes in the minimum cost combination in agriculture sketched so far have been "directed towards the necessity of increasing the efficiency of human labour by increasing the use of labour saving machinery and vehicles (mechanization). With the help of this equipment it is possible to open new areas to cultivation. The agricultural machinery industry is the bearer of progress in the development of new cultivation techniques. In turn its progress is favourably influenced by the expansion of the areas under cultivation and the increasing specialization of farms. Output per unit of area where agriculture is labour extensive . . ." is still relatively low. Agriculture is at the mechanization stage.

*Industrialized countries*

England may be taken as an example of industrialized countries at stage 4. As a result of the high degree of industrialization land and labour have grown still more expensive and scarce, while capital goods have become still less expensive. It now becomes urgent to combine high labour productivity with high land productivity, even if this can only be achieved by an exceptionally high input of capital and a resulting drop in capital productivity. The agricultural system must be intensive, particularly capital intensive.

At this stage, however, the increase in the capital input takes different forms. Herlemann says, "The improvement of the purchasing power ratio between agricultural products and production equipment produced by industry now makes it possible to increase the use of equipment and materials which increase yields in combination with better quality seed, improved crop rotation, better tillage, seed dressings and so forth. The agricultural machinery industry, increasingly concerned with the manufacture of labour saving machines and equipment (drilling machines, mechanical hoes, fertilizer distributers, cultivators, seed cleaning and dressing equipment, etc.), and the chemical industry (fertilizers, weed killers, seed dressings, vaccines, preservatives, etc.) become increasingly important to agricultural production." Because of the rising cost of land, the use of equipment and production materials which economise on the use of land comes to take its place beside the use of labour saving capital goods.

## Development of an overpopulated agricultural country into an industrialized country

In densely populated areas of the old world, where there were over 60 inhabitants per 100 ha of farmland the reverse process with regard to capital input gradually took place. In these countries land saving capital goods were more important at first and labour saving equipment was introduced only at a later stage. Here the intensification phase preceded the mechanization phase.

*Overpopulated agricultural countries*

The final type of development is that of an overpopulated agricultural country. Countries at stage 6 include India, China, Java and densely populated African countries such as Morocco, Ghana, Nigeria, Uganda, Rwanda and Burundi. Land is scarce and therefore expensive, labour is abundant and therefore cheap and capital goods are expensive because of the lack of industrialization. As a result one must accept low labour productivity but must have a high land and capital productivity. These economic targets are reached through capital extensive, labour intensive and on the whole moderately intensive agricultural production methods. The economic conditions in these countries at stage 6 are the most unfavourable because "the absence or extreme slowness of the process of industrialization, lagging far behind the natural growth of the population, leads to increasing overpopulation in rural areas with all its unfavourable effects on the productivity of labour and the standard of living. In spite of a high intensity of labour and a predominantly vegetarian diet, the domestic agricultural output guarantees only a minimum calorie supply. An excessively high birth rate, high fluctuations in crop yields and an inadequate import capacity increase the difficulties of supply which find their expression in periodical famines and epidemics. Since the prospects for successfully limiting the number of births artificially are dim and emigration possibilities rather restricted the ultimate solution lies in a well planned policy of industrialization. Owing to the dearth of domestic capital resources this must be supported by foreign capital. This solution is particularly to be desired because an increasing division of labour within the economy will not only raise the productivity of agriculture and create new jobs but will also slow down the natural growth of the population despite increasing national prosperity and an increased average life expectancy" (Herlemann 1954).

### Agro-industrial countries

The beginning of the process of industrialization brings us to the agro-industrial countries at stage 5. These include most industrialized countries of western Europe and Japan. Land prices are still high, wages are on the increase, but the prices of capital goods are lower. To take these cost conditions into account high productivity of land must be coupled with increasing productivity of labour allowing for diminishing productivity of capital. Agricultural production is thus based on increasing intensity of capital and decreasing intensity of labour.

"The prevailing scarcity of land coupled with a still adequate supply of labour seems to justify all expenditures that serve to increase the productivity of land. Such conditions explain Germany's leading position in the development of techniques for producing synthetic mineral fertilizers. The intensity of agriculture in West European countries, which is based on relatively high inputs of labour and increasingly frequent use of yield-increasing production materials, is expressed in terms of relatively high crop yields per unit of land and a relatively low productivity of agricultural labour compared to overseas industrialized agricultural countries."

### Industrialized countries

It is only "when, in the wake of a progressive division of labour in the national economy, agricultural labour becomes increasingly scarce and costly that agriculturists in these countries are forced to enhance the productivity of labour by employing more machinery on farms. This trend is favoured by the fact that the agricultural machinery industry, which originally catered mainly for large scale farms, has now largely adjusted its production to the needs of the system of family farms which prevails in Europe. They supply farmers with power driven implements, pneumatic-tyred all-purpose tractors, multipurpose implements, implement carriers and mechanized equipment for whole sequences of operations. The progress made by the agricultural machinery industry seems to remove the fear that the economically imperative need to mechanize might mean the end of the independent family farm and necessarily lead to collectivization. Most countries of western Europe, including Denmark and Holland, which export agricultural products, are at present in the mechanization phase (transition from stage 5 to stage 4)." (Herlemann 1954). The aim of economic development—whether in an overpopulated or a thinly populated agricultural country—is in every instance stage 4, which is characterized by a high input of capital designed to obtain a high productivity of land and labour.

The ideas so far come from Herlemann (1954, 1961), who later on developed and substantiated them to form a theory which I believe to be extremely helpful in understanding the position of the developing countries.

## Theory and Practice in Agriculture

The following sections go on to examine whether day-to-day agricultural practice supports this theory. It must, however, be stressed once again that a minimum cost combination is not a static concept but on the contrary a highly dynamic one. In every country it is subject to certain trends in development which differ in every case, but all culminate in the same point, the industrialized state. This has the following characteristics. A high degree of industrialization makes for a high level of wages and for a shortage of labour resulting in the need for a high productivity of labour. A high population density in relation to farmland also requires a high land productivity. This is feasible as soon as the level of agricultural prices increases with the growing purchasing power of the consumer. The productivity of land and labour must be increased simultaneously. This can only happen, however, if the cost of capital inputs is low. This makes it possible to increase labour productivity decisively through mechanization and to increase land productivity through inten- sification. Human labour is replaced by machines and land by yield-increasing equipment and farm inputs. The ultimate combination of factors of production is thus characterized by the economical use of land, the economical employment of labour and a very generous use of every form of purchased capital input. Agriculture is capital intensive.

### Differences in the minimum cost combination of countries at different stages of industrial development

A comparison between countries at different stages of industrial development will show considerable divergencies in the minimum cost combination in agriculture.

In Table 2 representative countries are taken as examples to demonstrate the change in the price relationships of factors of production in the course of industrialization, the reaction of agriculture to this change by a recombination of factors of production, and the consequent development in their productivity. Unfor- tunately I am not in a position to supply suitable data on African countries other than Zaire because so far only the South African Republic has reached a higher stage of industrialization. Since most African countries must still be regarded as thinly populated, they would belong to the group in columns 1–4 which show the development of a thinly populated agricultural country into an industrialized country. The last column (German Federal Republic) should be ignored for the moment.

The countries in columns 1–4, Zaire, Brazil, Argentina and the USA, are classified according to their stage of industrialization. The industrialization factor, defined as the value of industrial production as a multiple of the value of agricultural production, grows from 0.5 in Zaire to 9.2 in the USA. In the process of industrialization the percentage of the total working population employed in

Table 2    Price and quantity ratios and productivity of production factors in agriculture in the course of industrialization.

| Industrialization factor/country | 0.5/Zaire | 0.9/Brazil | 1.5/Argentina | 9.2/USA | 9.5/GFR |
|---|---|---|---|---|---|
| General economic situation: | | | | | |
| Inhabitants per 100 ha farmland | 29 | 51 | 14 | 41 | 389 |
| % of total working population in agriculture | 85 | 57 | 25 | 9 | 14 |
| % contributed by agriculture to GDP | 40 | 27 | 20 | 4 | 6 |
| Per-caput income (US $) | 60 | 310 | 508 | 2330 | 1200 |
| Cost of production factors (in grain units) | | | | | |
| Rent per ha of farmland | 1.2 | 1.5 | 1.6 | 2.5 | 7 |
| Monthly wages | 1.9 | 4.2 | 7.9 | 22 | 5.4 |
| 100 full fertilizer units* | 12.8 | 11.2 | 21.8 | 9.4 | 5 |
| Tractors (25–34 HP) | 995 | 879 | 726 | 376 | 236 |
| Input of production factors: | | | | | |
| Full-time workers (2400 hours per year) per 100 ha farmland | – | 8.3 | 1 | 1.3 | 18.9 |
| Mineral fertilizers (in 100 kg pure nutrient) per 100 ha farmland | – | 1.6 | 0.1 | 15.3 | 173 |
| Number of tractors per 100 ha farmland | – | 0 | 0.1 | 0.1 | 6 |
| Mineral fertilizers (in 100 kg pure nutrient) per 100 workers | – | 19 | 9 | 1200 | 920 |
| Number of tractors per 100 workers | – | 1 | 8 | 80 | 30 |
| Net productivity of production factors (three-year average): | | | | | |
| Land (grain units per ha) | – | 3.1 | 4 | 8 | 27.7 |
| Labour (grain units per worker) | – | 40 | 339 | 514 | 244 |

* 1 full fertilizer unit = 1 kg N + 1 kg $P_2O_5$ + 1.5 kg $K_2O$

Source: H. Mühl (1967).

agriculture falls from 85%, to 9%, and agriculture's percentage contribution to gross domestic product falls from 40%, to 4%. Per caput income rises from US $60 to US $2,330 as a result of the development of productivity of the economy as a whole.

One can now see how the price relationship of factors of production changes in the course of industrialization and how, as a consequence, the relationship between the quantities of different production factors used also changes. Rent per ha of farmland almost doubles, and wages are multiplied nearly twelvefold. Our theory suggests the intensity of labour should therefore decrease. The table shows that the employment of full time workers (2400 working hours per year) per ha of farmland does in fact decrease from 8.3 to 1.3.

A further consequence of industrialization is a reduction in prices of all capital goods in agriculture. This applies both to mineral fertilizers and tractors. The logical consequence is an increase in the use of these purchased inputs. The examples refer to thinly populated countries and the facts match well with the theory. Up to Argentina's stage of development the use of labour saving tractors increases whereas in the subsequent stages of development the use of tractors levels off and the increased investment in purchased capital inputs shifts to land saving mineral fertilizers.

When the capital input is related to full time agricultural workers three facts emerge:

(1) The capital input grows considerably in the course of development because the cost of capital decreases while the cost of labour increases. This means that the marginal productivity of labour must be progressively increased at the expense of the marginal productivity of capital.

(2) Up to Argentina's stage of development the use of mineral fertilizers is negligible while the use of tractors increases eightfold. This is because land is still plentiful although labour is becoming increasingly scarce and costly as industrialization progresses.

(3) From the stage of development in Argentina to that in the USA the use of mineral fertilizers increases far more rapidly than the use of tractors. This is because although it is still necessary to economize on the use of labour it is even more necessary to economise on land which is becoming increasingly costly, by using production materials to increase crop yields.

The result of this recombination of factors of production is that the net productivity of the land has become two and a half times as great while the productivity of labour has become 13 times as great. This increase in productivity is the characteristic feature of industrial development and a primary increase in the productivity of labour is the characteristic feature of all thinly populated countries.

## Differences in the minimum cost combination of countries with varying population densities (USA/Federal Republic of Germany)

According to the theory set out above, the minimum cost combination in agriculture depends on the one hand on the degree of industrialization and on the other on the population density. The comparison between Brazil, Argentina and the USA showed that development was largely influenced by the degree of industrialization because these countries are mostly thinly populated (their population density varies only between 14 and 51 inhabitants per 100 ha of farmland). On the other hand very considerable differences can be seen in the minimum cost combination of countries at the same stage of industrialization with widely differing population densities. This can be seen by comparing the USA and the German Federal Republic, whose factors of industrialization at 9.2 and 9.5 respectively are practically the same although the population density of the German Federal Republic is almost 10 times that of the USA.

At its present stage the USA must be regarded as an industrialized agricultural country (Table 1, stage 3). It has developed from a thinly populated agricultural country. The German Federal Republic at its present stage of development must be classified as an agroindustrial country (Table 1, stage 5). It has developed from an overpopulated agricultural country. Neither country has yet reached the highest possible stage of industrialization such as may be found, for example, in Great Britain. However, the various stages of capital input in both countries demonstrate the theory well. Thus in the thinly populated USA the mechanization phase preceded the intensification phase, whereas in the densely populated German Federal Republic the intensification phase came first and was later followed by the mechanization phase.

How does the minimum cost combination in agriculture in these two countries differ? As Table 2 illustrates, the relative prices of the production factors are again decisive. The relative costs of land and labour are diametrically opposite in the two countries. While rents are three times as high in the German Federal Republic as in the USA, wages in the USA are four times as high as those in the German Federal Republic. The prices of capital goods are relatively low in both countries compared to those in developing countries. These differences in the ratios of prices result in the use of different proportions of the production factors.

*Intensity*

The German farmer has to pay three times the price the American farmer pays for land; he therefore has to use his land more economically than labour and capital goods so as to obtain higher marginal yields per ha than the American farmer. Thus Table 2 shows that for every 100 ha of farmland the Federal German farmer uses almost 15 times the input of labour, 11 times the amount of mineral fertilizer and 55 times the number of tractors used by the farmer in the USA. Thus Federal German agriculture is much more intensive than that of the USA. This more intensive agricultural production method results in a considerably higher land productivity. The net yield of the land in grain units per ha is three and a half times that of the USA.

*Input structure*

The best combination of labour and capital in either the USA or the German Federal Republic can be seen from a comparison of the relative costs of labour and of capital goods invested in agriculture. Measured in terms of their barter value against agricultural products most farm equipment and production materials are more expensive in the USA than they are in the German Federal Republic. The barter value of labour in terms of agricultural products is even more unfavourable in the USA. In 1962 one hour's wages cost the equivalent of 4.9 kg of wheat in Germany, 10 kg of wheat in Sweden and as much as 11.7 kg of wheat in the USA. The monthly wage of an agricultural worker in the USA is sufficient to purchase a great deal more equipment and production materials than the monthly wage of an agricultural worker in Germany. Compared to labour in the German Federal Republic labour in the USA is much more expensive than industrially produced equipment and production materials. The minimum cost combination of production factors will therefore result in a much higher utilisation of technical equipment in support of human labour in the USA than in the German Federal Republic. American farmers are forced to replace human labour by machines and within given limits by fertilizers and feeds to much greater extent than German farmers (see Table 2).

   The extremely low input of labour and comparatively high inputs of materials and machinery in American agriculture, always surprising to the German agriculturalist, are the result of the American wage level which is extremely high both in relative and absolute terms. This in turn is the cause of the very high labour productivity in American agriculture which results in net yields per worker (in grain units) three times as great as those in the German Federal Republic.

*Dynamic approach*

Specifically American agricultural methods of production are geared to a high productivity of labour whereas the specifically German methods are oriented towards high land productivity. The principle of minimum costs explains these differences in production methods as a sensible economic reaction to the specific price conditions in the two countries, in both of which farmers are seeking to achieve the same aim of maximum returns. Although it is true that in recent years American agriculturists have also been seeking to increase land productivity, the primary emphasis still remains firmly on labour productivity. On the other hand, in Germany increasing attention is now being given to the improvement of labour productivity while still maintaining high land productivity. The two countries will very slowly grow nearer together in their minimum cost combinations but only to the extent to which the relative prices of the production factors in the one country gradually approach those in the other. Only through such price adjustments will scope be gained for a corresponding recombination in the quantities of land, labour and capital which are used.

## Differences in the minimum cost combination of countries at different stages of industrialization and with different population densities

Differences in the minimum cost combinations in agriculture become particularly marked if two countries are compared which differ not only with regard to the degree of industrialization they have attained but also with regard to the density of their populations. Thailand and the USA are two countries that differ in both these respects. In Thailand there are 321 inhabitants per 100 ha of farmland and in the USA only 41. Thailand's industrialization factor is only 0.5, whereas that of the USA is 9.2. Under the theory in Table 1 Thailand is an overpopulated agricultural country at stage 6, while the USA is one of the thinly populated but heavily industrialized countries at stage 3.

The following section compares the methods of growing rice in both countries. This not only further confirms the theory but also shows that even within the same branch of agriculture radical changes in the combination of production factors may take place.

The initial relations between production factors and prices are as follows: In Thailand labour is cheap, but land and capital are expensive. In the USA labour is expensive, but land and capital are cheap. In Thailand the minumum cost combination for agriculture therefore requires high land and capital productivity involving the employment of large numbers of agricultural workers and in the USA it requires high labour productivity achieved through the extensive use of land and high inputs of capital.

It is possible to see from Table 3 that there are interesting differences between the methods of cultivating rice in both countries. The Thai family farm is only 0.75 to 4 ha of farmland, while the US family farm has 75 to 250 ha at its disposal for growing rice. In Thailand, where labour is abundant, land scarce and capital goods expensive, rice is cultivated as a root crop; in the USA, where wages are high and land and capital goods abundant and cheap, rice is cultivated as a grain crop. As a result one hectare of rice in Thailand requires 600 to 1200 working hours and in the USA only 20 to 30.

Table 3    Comparison of methods of growing rice in Thailand and the USA.

| Process | Thailand | USA |
|---|---|---|
| Initial price ratio of production factors | Labour cheap; land and capital expensive | Labour expensive; land and capital cheap |
| Size of a family farm | 0.75 to 4.0 ha | 75 to 250 ha |
| Labour required | 600 to 1200 hours/ha | 20 to 30 hours/ha |
| Ploughing method and speed | Digging stick and hoe 0.05 ha/per worker per day | Tractor 6 to 7.5 ha per worker per day |
| Cultivation method | Initial planting in seed beds; transplanting of shoots | Airplane seeding |
| Fertilizers and pest control measures | Hardly used | Airplane spraying by contractors |
| Irrigation | Scoop wheel and capstan driven by water buffalo | Large diesel pumps |
| Harvesting | Panicles cut by knife; drying in stooks, etc.; threshing by wooden sled drawn by water buffalo | Large combine harvesters, automotive, 10 km/ha; up to 6 m cutting width up to 10 000 kg/ha; the largest combines require servicing by only one man and harvest 300 to 400 ha per year |
| Cultivation method | As a root crop | As a grain crop |
| Yields 1975 | 1 770 kg/ha | 51 000 + kg/ha |
| 1974 Producer prices of rice | $0.099 kg | $0.230 |

In Thailand one worker can till only 0.05 ha per day with a digging stick or hoe. In the USA one worker can plough 6 to 7.5 ha per day by tractor. The plant stock in Thailand is started by the growing of seedlings in seedbeds and each individual plant is subsequently transplanted on to the rice field. In California, on the other hand, contractors are hired to carry out aerial seeding. Fertilizer application and pest control are hardly carried out at all in Thailand, while in the USA contractors are hired here too to undertake aerial spraying. In Thailand the rice fields are irrigated by means of scoop wheels and capstans driven by water buffalo. In the USA gigantic pumping plants with diesel motors are used for irrigation.

Harvesting methods are also totally different in the two countries. In Thailand the panicles are cut by knife or sickle, then dried in stooks and finally threshed by wooden sled drawn by water buffalo. In the USA large motorized combine harvesters are employed. These can work at a rate of 10 km/h, have a cutting width of up to 6 m. and are capable of threshing 10 000 kg per hour. The largest combine harvesters can be worked by a single man and can harvest 300 to 400 ha of rice per year.

B

## Summary

These examples are sufficient to demonstrate how the minimum cost combination in agriculture varies at different stages of national economic development. The theory may be summarized by saying that in agriculture it is necessary to distinguish the three basic production factors, land, labour and capital. From a technical point of view there is no single specific ratio in which these three factors have to be combined. Within wide limits they are interchangeable. From the economic point of view, however, there is only one single minimum cost combination of these three production factors for each stage of economic development. This is derived from the relationship between the factor prices (ground rent, wages, interest) and is the combination of factors which ensures that production is as cheap as possible.

In a farm run on the profit principle the marginal returns (measured in money) on each of the variable factors of production should be sufficient to cover the cost of the input of that factor. In addition the marginal productivity of each factor sinks if its input is continually increased relative to that of the other two factors. It follows that, other things being equal, the higher the price of a factor of production the more sparingly it must be used. Correspondingly the input of the cheapest of the three production factors should be quantitatively the greatest.

Progress in national economic development, which is generally achieved through increasing industrialization, leads to a situation in which capital goods gradually grow cheaper while wages are constantly on the increase. One of the results of this is that human labour must be increasingly supplemented by capital goods. The degree of availability of land determines which type of capital goods will be of primary importance in the course of this general economic development. In thinly populated agricultural countries the main concern is to replace manual labour by mechanical equipment. In the initial phases this is far more important than economizing on land. As a result labour saving capital goods are of prime importance and mechanization precedes intensification. Not until later, when agricultural land begins to grow scarce, will production materials which increase yields be used in increasing amounts.

The situation in overpopulated agricultural countries is entirely different. Here land becomes scarce more rapidly than labour, so that production materials which increase yields must first be given priority so as to economize in the use of land. Here intensification precedes mechanization. Mechanization begins later as labour in rural areas becomes increasingly scarce and expensive as a result of industrial progress.

In short the minimum cost combination in agriculture varies greatly from country to country and must change considerably from phase to phase in the course of national economic development. The criteria according to which the minimum cost combinations of different developing countries tend to change vary greatly, depending on whether countries are initially sparsely or densely populated. It also follows that in developing countries with economic structures which are different, the importance of technical progress will also vary. It is easy to see that thinly populated agricultural countries will benefit most from progress in mechanization. On the other hand overpopulated agricultural economies will benefit most from progress in the biological field.

# Bibliography

Aereboe, F. (1923) [General agricultural management studies.] Allgemeine landwirtschaftliche Betriebslehre. 6th edn. Berlin; Paul Parey 697pp.

Andreae, B. (1958) [Productivity zones in the agricultural region of North America.] Produktivitätszonen im Agrarraum von Nordamerika. *Agrarwirtschaft* 7, 75–80.

Andreae, B. (1968) [Minimal cost combination in agriculture in the course of economic development. Examples and verification of a theory by H.H.Herlemann.] Die Minimalkostencombination in der Landwirtschaft im Zuge der Volkswirtschaftlichen Entwicklung. Exemplifizierung und Verifizierung einer Theorie von H.H.Herlemann. *Berichte über Landwirtschaft* 46, 1–196.

Andreae, B. (1974) [Diversification and specialization of the farm sector in the tropics.] Diversifizierung und Spezialisierung der Farmwirtschaft im Tropenraum. *Berichte über Landwirtschaft* 52, 497–511.

Andreae, B. (1978) The minimum cost combination in agriculture. *Geo Journal* 2 (3), 203–214.

Andreae, B.; Greiser, E. (1978) [Structures of the German agricultural landscape.] Strukturen deutscher Agrarlandschaft. *Forschungen zur Deutschen Landeskunde* No.199, Ed.2.

Brinkmann, T. (1922) [The economics of the farm enterprise.] Die Ökonomik des landwirtschaftlichen Betriebes. In: Grundriss der Sozialökonomik, Part 7, Land- und forstwirtschaftliche Produktion, Versicherungswesen. Tübingen, German Federal Republic; J.C.B.Mohr (Paul Siebeck) 27–124.

Healey, D.T. (1964) Agricultural economics in some African countries. *International Journal of Agrarian Affairs* 4 (4), 250–286.

Herlemann, H.H. (1954) [Stages in the technical development of agriculture.] Technisierungsstufen der Landwirtschaft. *Berichte über Landwirtschaft* 32 335–342.

Herlemann, H.H. (1961) [Principles of agricultural policy.] Grundlagen der Agrarpolitik. In: Kade, G. (*Ed*) Vahlens Handbücher der Wirtschafts- und Sozialwissenschaften. Berlin and Frankfurt/Main, German Federal Republic; Franz Vahlen GmbH 191pp.

Mühl, H. (1967) [The combination of factors of production in agriculture in the course of economic and technical development with particular reference to inputs of mineral fertilizers.] Über die Kombination der Produktionsfaktoren in der Landwirtschaft im zuge der wirtschaftlich-technischen Entwicklung unter besonderer Berücksichtigung des Mineraldüngereinsatzes. Dissertation, Technical University, Berlin 169pp.

# Chapter 3

# Typology of the Climatic Zones and Farming Systems in the Tropics and Sub-tropics

Over the world, agricultural zones are on the whole determined mainly by climate and the stage of economic development. However, an initial survey of farming regions in the tropics can be based on climatic factors alone, since the economies of developing countries which are the only ones in these lower latitudes do not, so far, differ greatly.

Although the primary concern is the connection between farming regions and climate, population density is also considered in this chapter as an additional causal factor. The interaction between natural and cultivated vegetation is also particularly interesting.

To illustrate the climatically determined difference between the farming regions of the tropics, it is necessary first to identify both the types of climate and the types of farming.

## Climatic Zones and Regions

The region which spans the equator and extends between the two tropics includes both the most arid and the most humid climatic belts on the earth. Fig. 3 gives a rough indication of the position of the tropical climatic zones within the world's climatic belts. A newer classification, based on climographs, is given by Walter (1977).

For the purpose of this study, however, a stricter classification of tropical climates is needed. Optimally this should be one based on vegetation formations. With increasing proximity to the equator and with rising humidity, six climatic zones can be distinguished.

The zones shown in Fig. 4 are represented by the following six vegetation forms of which some examples are given:

1. The dry-hot desert, with only slight and sporadic rainfall, which is mostly beyond the aridity limit for human habitation: (e.g., the Sahara, Kalahari and Central Australian Deserts).
2. The arid semi-desert, partly inside and partly outside the limit for grazing livestock. This can only be used occasionally by nomads and by hunters and gatherers (e.g., large parts of Egypt, Libya, Mauretania).
3. The semi-arid shrubland-steppe (shrub, salt steppe), with two to four humid months, which extends up to the agronomic aridity limit. Apart from a very little irrigated farming and some dry farming the steppe is used almost exclusively for extensive pasture—i.e. for ranching in the New World and often still by nomads in the Old World (e.g. large parts of Senegal, Upper Volta, South West Africa).
4. The dry savannah, with four to six humid months, situated between the agronomic and climatic aridity limits and thus capable of supporting rainfed

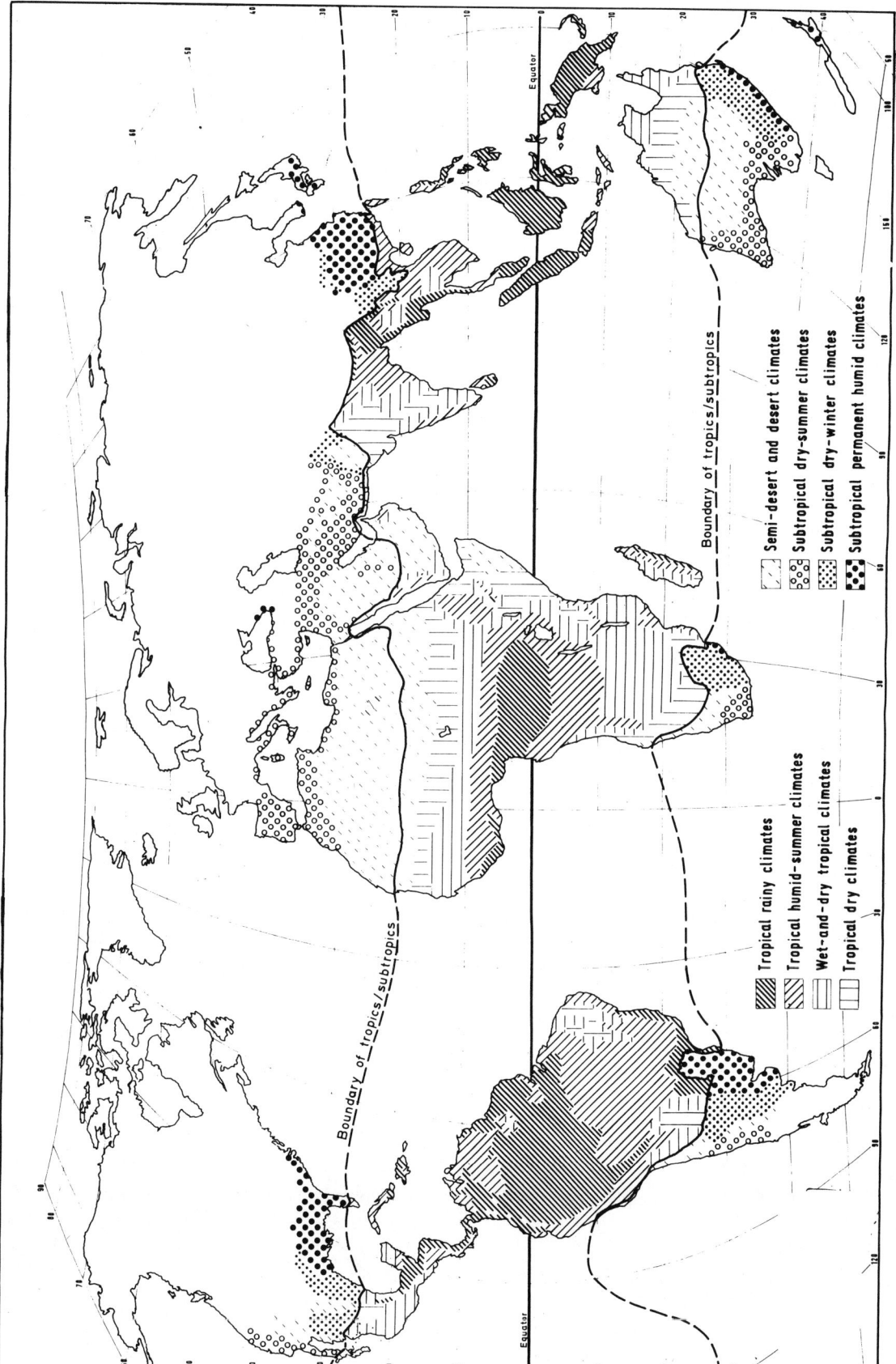

Key (legend):

**Semi-desert and desert climates**
**Subtropical dry-summer climates**
**Subtropical dry-winter climates**
**Subtropical permanent humid climates**

**Tropical rainy climates**
**Tropical humid-summer climates**
**Wet-and-dry tropical climates**
**Tropical dry climates**

Boundary of tropics/subtropics

Equator

Fig. 3
Seasonal climates of the tropics and sub-tropics. (Source: Troll & Paffen 1963)

Fig. 4
Climatic zones of the tropics

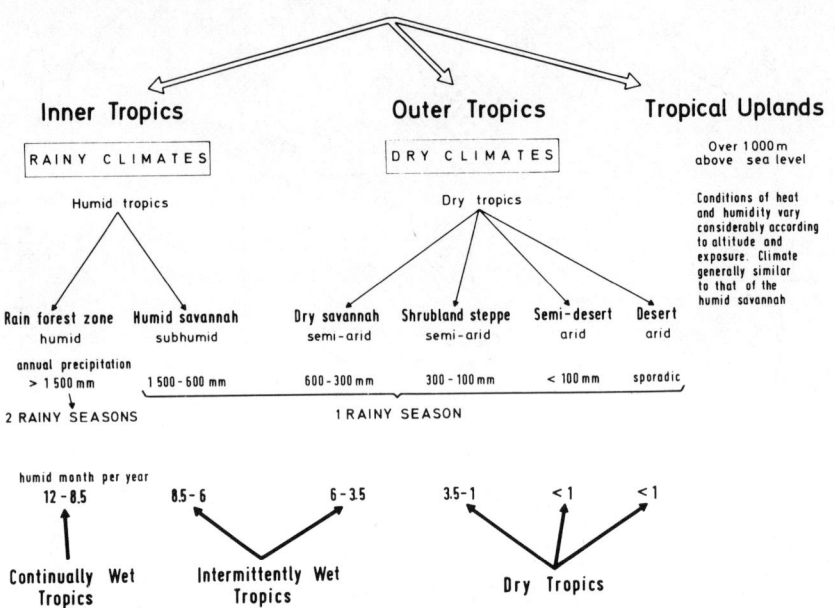

farming. The grassland which occurs is dry steppe, and the woodland is dry forest which is green in the rainy season (e.g. Miombo). A short rainy season is followed by a long dry season (e.g. large parts of Chad, Tanzania, Somalia).

5. The sub-humid wet savanna, with seven to nine humid months, situated between the climatic aridity limit and the humidity limit for pasture farming. Here the rainy season is longer, the dry season shorter and the atmospheric humidity higher. The yearly rainfall varies from approximately 800 mm to 1500 mm. The grassland which occurs is high-grass savannah with tropical river forest, and the woodland is monsoon forest. This is the first climatic zone where even the small rivers contain water all the year round: in zones 1 to 4 they dry up soon after the rainy season ends (e.g. large parts of North Brazil, Thailand, Ghana).

6. The tropical rain forest, immediately on the equator and with the wettest climate. There are two rainy seasons with a total of at least 1500 mm. annual rainfall, the mean temperature is 25° to 28°C and varies little throughout the year, and the humidity is rarely below 90%. This leads to a thoroughly humid, continually wet type of climate giving rise to evergreen tropical forest (e.g. large parts of Indonesia and of the Congo and Amazon basins).

7. In addition there are the tropical uplands (over 1000 m above sea level) with very different types of vegetation. However, they can be classified partially, or even wholly, with the above groups especially with the dry and humid savannahs. Belonging to this group are large parts of East Africa, where agriculture has in some regions reached a remarkable stage of development because of the influence of the upland climate, efficient seaports and foreign populations, particularly the white settlers.

## Farming Types and Systems

The classification of types of agricultural system in the tropics follows approximately the same progression from the dry to the rainy climates. Only four broad types are met with:

1. Pasture farming. Up to now this has been completely extensive in the tropics, since there is no cultivation of any kind. The population simply makes use of anything nature will grow without human intervention and does this by the use of undemanding grazing animals. Livestock housing is completely unknown, as is the conservation of fodder generally.
2. Rain fed farming, also called dryland farming, involves the cultivation of short-lived crops on the basis of the natural rainfall, i.e., without irrigation. The problem of soil fertility often makes it necessary to intersperse the cultivation of useful crops with bare fallow, grass, shrub or even forest fallow. In extreme cases (e.g. the forest burning system of shifting cultivation) only 15 to 25% of the available land can be cropped and harvested at any one time.
3. Irrigated farming provides short-lived crops with an artificially increased water supply, either throughout the year or simply in the dry season, so that crop production is often continued through the whole year. With some crops

Fig. 5
Humid and arid months, vegetation belts and farming systems. (Source: Uhlig 1965, p.3)

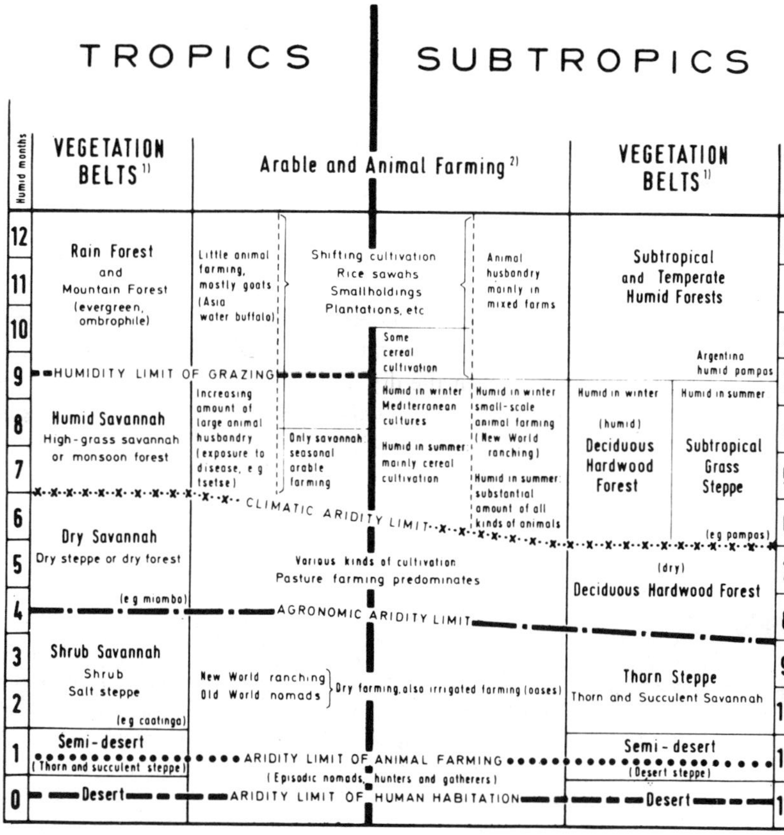


1) And vegetation formations — 2) Only some typical cropping pattern and animal activities

it is then possible to obtain two or even three harvests per year from the same land.

4. Perennial crop farming. This involves growth cycles of a few to a great many years, often several decades, as with the cultivation of coconuts, oil palms, rubber, citrus fruits, coffee, cocoa and tea. The continuous land cover and the technical processing of the harvested crop are important characteristics of this type of farming. Most farms practise intensive cultivation. There is a widespread tendency towards monoculture and to the plantation scale of farming.

Fig. 5 shows which types of arable farming and animal farming are native to the climatically determined, natural vegetation belts of the tropics. To gain an overall impression of the vegetational characteristics and agricultural geography of the

Fig. 6
Farming systems in the climatic zones of the tropics

| Climax vegetation | Humid months | Ranching | Rain-fed farming | Irrigated farming | Tree and shrub cultivation |
|---|---|---|---|---|---|
| 1. Deserts | — | | | | |
| 2. Semi-deserts | 1 | ● | | | |
| 3. Shrubland steppes | 1 - 4 | ● ● ● | ● | ● | |
| 4. Dry savannahs | 4 - 6 | ● ● ● | ● ● | ● ● | ● |
| 5. Wet savannahs | 7 - 9 | ● | ● ● ● | ● ● ● | ● ● |
| 6. Rain forests | 10 - 12 | | ● ● | ● ● | ● ● ● |

● = infrequent          ● ● = moderately frequent          ● ● ● = widespread

area, it is especially important to obseve that in Fig. 5: (1) the aridity limit for habitation runs through the desert; (2) the aridity limit for animal farming passes through the semi-desert; (3) the agronomic aridity limit (limit for rain fed farming) divides the shrubland steppe and the dry savannah; (4) the climatic aridity limit runs between the dry and humid savannahs; and (5) the humidity limit for pasture farming is identical with the contact zone between the humid savannah and the rain forest. Fig. 6 attempts to give an overall picture of the distribution of the four systems of agriculture within the six climatic zones and Table 4 shows some developing countries in the different climatic zones.

## Bibliography

Andreae, B. (1965) [Land fertility in the tropics. Utilization and maintenance. Farm management considerations for work in developing countries.] Die Bodenfruchtbarkeit in den Tropen. Nutzbarmachung und Erhaltung. Betriebswirtschaftliche Überlegungen für die Arbeit in Entwicklungsländern. Hamburg, German Federal Republic; Paul Parey 124pp.

Andreae, B. (1971) [Dryland farming in the tropics. Climatic zones, land forms, associated crops and development aims.] Trockenfeldbau in den Tropen. Klimazonen, Grundformen, Kulturpflanzengemeinschaften und Entwicklungsziele. *Landwirt im Ausland* 5, 33–35.

Table 4 Developing countries in different climatic zones.

| Climatic zone | Vegetation belt | Developing countries (examples) | Population | | Economically active population in agriculture in %, 1976 | Per capita gross domestic product in $, 1975 | Main agricultural export products in %, 1975 |
|---|---|---|---|---|---|---|---|
| | | | Density per km² 1975 | Annual rate of increase 1970–75, % | | | |
| Inner tropics | Rain forest | Zaire | 11 | 2.8 | 76.3 | 146[2] | . |
| | | Indonesia | 20 | 2.9 | 62.0 | 126[3] | 41.6 rubber; 21.4 coffee, tea etc., |
| | Humid savannah | Ghana | 41 | 2.7 | 53.8 | 257[1] | 98.9 coffee, tea, cocoa |
| | | Tanzania | 16 | 2.9 | 83.1 | 170 | 42.8 coffee, tea etc., 28.7 textile fibres |
| | | Thailand | 81 | 2.9 | 77.2 | 342 | 40.5 cereals, 20.3 sugar |
| Tropical highlands | Varying considerably | Ethiopia | 23 | 2.6 | 81.2 | 98[2] | 33.2 coffee, tea, cocoa 17.8 oilseeds |
| | | Peru | 3 | 2.6 | 40.3 | 546 | 66.1 sugar, 11.5 coffee etc. |
| Outer tropics | Dry savannah | Senegal | 21 | 2.2 | 76.6 | 294[2] | 67.9 vegetable-oils 18.5 animal feedstuffs |
| | | Chad | 3 | 2.1 | 86.6 | 74[1] | 52.9 textile fibres 35.3 live animals |
| | Shrub savannah | Mauritania | 1 | 2.6 | 84.8 | 232[3] | . |
| | | Namibia | 1 | 2.9 | 51.3 | . | . |
| | Semi-desert, desert | Saudi Arabia | 4 | 3.0 | 62.5 | 3220[2] | 34.4 fruit and vegetables 19.2 tobacco |
| Subtropics | Varying considerably | Pakistan | 87 | 3.0 | 55.5 | 368 | 50.7 cereals, 34.8 textile fibres |
| | | Morocco | 39 | . | 53.4 | 426 | 76.8 fruits and vegetables |
| | | Uruguay | 17 | 1.2 | 13.2 | 1153 | |
| For comparison with developing countries | | United Kingdom | 229 | 0.2 | 2.3 | 4089 | 27.8 beverages, 10.8 cereals |

[1] 1970; [2] 1974; [3] 1973.
Sources: FAO (1977a) pp.61ff, (1977b) pp. 297ff, U.N. Statistical Office (1977) pp. 297 ff.

Andreae, B. (1977) [Agricultural geography.] Agrargeographie. Berlin; New York, USA; Walter de Gruyter 332pp. [English language edition forthcoming, New York 1980.]

Andreae, B. (1977) [Agricultural systems.] Agrarsysteme. In Albers, W.; Burn, K.E.; Dürr, E. (*Ed*) Handwörterbuch der Wirtschaftswissenschaft, Vol.1 Göttingen, German Federal Republic; Vandenhoeck & Ruprecht 155–169.

Andreae, B. (1977) Farming regions in the tropics. In: Handbook of vegetation science, Part 13, (Edited by W.Krause). The Hague, Netherlands; Dr.W.Junk B.V. 161–209.

Andreae, B. (1978) [Agricultural regions under local stress.] Agrarregionen unter Standortstress. Kiel, German Federal Republic; Verlag Ferdinand Hirt 78pp. [Geocolleg.No.6].

Andreae, B. (1976) [Stages in planning of small farms at the generation change.] Planungsstufen von Kleinfarmen im Generationswechsel. *Geographische Rundschau* 28, 304–308.

Baren, F.A. van (1964) [Physical-geographical aspects of arid and semi-arid zones.] Fysisch-geografische aspecten van de aride en semi-aride gebieden. *Tijdschrift van het Koninklijk Nederlands Aardrijkskundig Genootschap* 81, 182–195.

Bennett, M.K. (1960) A world map of food-crop climates. *Food Research Institute Studies* 1 (3), 285–295.

Bertalanffy, L. von (1973) General systems theory. Ed.4 New York, USA; George Braziller 289pp.

Best, R. (1962) Production factors in the tropics. *Netherlands Journal of Agricultural Science* 10, 347 et seq.

Boughey, A.S. (1957) The physiognomic delimitation of West African vegetation. *West African Science Association Journal* 3, 148 et seq.

Brandt, H. (1971) [The organization of peasant farms under the influence of the development of an industrial town: the case of Jinja, Uganda.] Die Organisation bäuerlicher Betriebe unter dem Einfluss der Entwicklung einer Industriestadt: Der Fall Jinja/Uganda. *Zeitschrift für Ausländische Landwirtschaft, Materialsammlung* 154pp. + 54pp.

Cabot, J. (1961) [In Chad, the problem of the Koros department of Logone: the example of the Sar plateau.] Au Tchad, le problème des Koros départment du Logone: l'exemple du plateau de Sar. *Annales de Géographie* 70, 621–633.

Chaves, V.L.F. (1963) [Geographical handbook for the use of soil and water conservation technicians.] Manual de geografia para el uso de los tecnicos de conservacion de suelos y aguas. *Revista Geografica, Instituto Pan-Americano de Geografia e Historia* 31, 81 et seq.

Daveau, S.; Ribeiro, O. (1973) [The humid inter-tropical zone.] La zone intertropicale humide. Paris, France; Librairie Armand Colin 275pp.

Douglas, H.K.L. (1957) Climate and economic development in the tropics. New York, USA.

Duckham, A.; Masefield, G.B. (1970) Farming systems of the world. London, UK; Chatto & Windus 542pp.

Duvigneaud, P. (1953) [The Savannahs of the lower Congo.] Les savanes du Bas-Congo. Liège, Belgium; Lejeunia 192pp.

Edwards, D.C. (1951) The vegetation in relation to soil and water conservation in East Africa. *Commonwealth Bureau of Pastures and Field Crops Bulletin* 41, 28–43.

Eyre, S.R. (1963) Vegetation and soils. A world picture. London, UK; Edward Arnold (Publishers) Ltd. 324pp.

Fao Production Yearbook (1970) No.23.

Fao Production Yearbook (1977a) No.30.

Fao Trade Yearbook (1977b) No.30.

Faucher, D. (1949) [Agricultural geography. Types of crops.] Géographie agraire. Types de cultures. Paris, France; Librairie de Médicis, Editions M.Th.Génin 382pp.

Fickendey, E. (1941) [Indigenous crops and plantations.] Eingeborenenkultur und Plantage. *Mitteilungen der Gruppe Deutscher Kolonialwirtschaftlicher Unternehmungen* No.4.

Finch, A. (1963) [Tropical soils. Introduction to principles of soil science in tropical and sub-tropical agriculture.] Tropische Böden. Einführung in die bodenkundlichen Grundlagen tropischer und subtropischer Landwirtschaft. Hamburg, German Federal Republic; Paul Parey 188pp.

Franke, G. et al. (1967) [Useful crops in the tropics and sub-tropics.] Nutzpflanzen der Tropen und Subtropen, Vols. I and II. Leipzig, German Democratic Republic; S. Hirzel Verlag 324pp.; 421pp.

Gregor, H.F. (1970) Geography of agriculture. Themes of research. London, UK; Englewood Cliffs 181pp.

Grigg, D.B. (1974) The agricultural systems of the world. An evolutionary approach. London, UK; New York, USA; Melbourne, Australia; Cambridge University Press 358pp.

Hasselmann, K.H. (1971) [Field book notes in Busoga 1966-1967.] Feldbuchaufzeichnungen in Busoga 1966-1967. (Excerpts from an unpublished manuscript, Berlin, cited from H.Brandt see above.

Hasselo, H.N. (1961) The soils of the lower eastern slopes of the Cameroon mountain and their suitability for various perennial crops. Wageningen, Netherlands; H.Veenman en Zonen 67pp.

Healy, D.T. (1964) Agricultural economics in some African countries. *International Journal of Agrarian Affairs* 4 (4), 250–286.

Holm, H.M. (1956) The agricultural economy of Ethiopia. Washington, USA; Foreign Agriculture Service, US Department of Agriculture 44pp.

Hueck, K. (1961) [Boundaries, ecological characteristics, types of location and economic importance of the rain forest area on Lake Maracaibo.] Grenzen, ökologische Merkmale, Standortstypen und wirtschaftliche Bedeutung des Regenwaldgebietes am Maracaibo-See. *Die Erde* 92, 193–204.

Huffnagel, F.R. (1961) Agriculture in Ethiopia. Rome, Italy; FAO 484pp.

Irvine, F.R. (1958) A text-book of West-African agriculture. London, UK; Oxford University Press 367pp.

Jaeger, F. (1946) [Climatic boundaries in arable farming.] Die Klimatischen Grenzen des Ackerbaues. *Denkschriften der Schweizerischen Naturforschenden Gesellschaft* No.76 48pp.

Kirsch, W. (1974) [Farm management studies: systems, decisions and methods.] Betriebswirtschaftslehre: Systeme, Entscheidungen, Methoden. Wiesbaden, German Federal Republic; Betriebswirtschaftlicher Verlag Dr.Th. Gabler 317pp.

Könnecke, G. (1967) [Crop rotation.] Fruchtfolgen. Berlin; VEB Deutscher Landwirtschaftsverlag 335pp.

Lauer, W. et al (1952) [Studies on the sciences of climate and vegetation in the tropics.] Studien zur Klima- und Vegetationskunde der Tropen. *Bonner Geographische Abhandlungen* No.9 182pp.

Lauer, W. (1956) [Vegetation, land use and agricultural potential in El Salvador (Central America).] Vegetation, Landnutzung und Agrarpotential in El Salvador (Zentralamerika). *Schriften des Geographischen Instituts der Universität Kiel* 16 (1), 98pp.

Lee, D.H.H. (1957) Climate and economic development in the tropics. New York, USA; Harper & Brothers 182pp.

Manshard, W. (1961a) [A suggested classification and nomenclature of forms of vegetation in Africa south of the Sahara.] Ein Vorschlag zur Gliederung und Benennung von Vegetationsformationen in Afrika südlich der Sahara. In: Meynen, E. (*Ed*) Geographisches Taschenbuch 1960/61. Wiesbaden, German Federal Republic; Franz Steiner Verlag 454–463.

Manshard, W. (1961b) [The geographical basis of the economy of Ghana with special reference to agricultural development.] Die geographischen Grundlagen der Wirtschaft Ghanas unter besonderer Berücksichtigung der agrarischen Entwicklung. *Beiträge zur Länderkunde Afrikas, Sonderfolge der Kölner Geographischen Arbeiten* No.1 308pp.

Manshard, W. (1974) Tropical agriculture. London, UK; New York, USA; Longman 226pp.

Mare, P.H. le (1959) Soil fertility studies in three areas of Tanganyika. *Empire Journal of Experimental Agriculture* 27, 191 et seq.

Masefield, G.B. (1948) The life of perennial crops. *East African Agricultural Journal of Kenya*.

McMaster, D.N. (1962) A subsistence crop geography of Uganda. *Occasional Papers, World Land Use Survey* No.2.

Mohr, E.C.J.; Baren, F.A. van (1959) Tropical soils. A critical study of soil genesis as related to climate, rock and vegetation. The Hague, Netherlands; London, UK; New York, USA; N.V. Uitgeverij W. van Hoeve 498pp.

Mukerjee, H.N. (1962) Problems of soil of the paddy fields. Transactions International Soil Conference New Zealand, Session C.13.

Niederstucke, H. (1970) [Forms of land use in the tropical highlands.] Bodennutzungsformen in tropischen Höhenlagen. *Landwirt im Ausland* 4, 74–76.

Nitz, H.J. (1973) [Systems of cultivation, common field systems and settlement systems.] Anbausysteme, Zelgenwirtschaften und Siedlungssysteme. In: Vergleichende Kulturgeographie der Hochgebirge des südlichen Asien. Wiesbaden, German Federal Republic; Franz Steiner Verlag 1–9 [Erdwissenschaftliche Forschung No.1].

Öbst, E. (1932) Classifying tropical zones by the crops they produce.] Die Gliederung der Tropenzone nach ihren Pflanzenerzeugnissen. Festschrift für Carl Uhlig. Öhringen, German Federal Republic; Verlag der Hohenloheschen Buchhandlung F.Rau 345pp.

Öbst, E (1965) [General economic and transport geography.] Allgemeine Wirtschafts- und Verkehrsgeographie. In: Öbst, E. (*Ed*) Lehrbuch der allgemeinen Geographie Vol.III, Ed.3, revised and expanded. Berlin; Walter de Gruyter 698pp.

Ochse, J.J.; Soule, M.J., Jr.; Dijkman, M.J.; Wehlburg, C. (1961) Tropical and subtropical agriculture. 2 vols. New York, USA; MacMillan 1446pp.

Otremba, E. (1960) [General agricultural and industrial geography.] Allgemeine Agrar- und Industriegeographie. In: Erde und Weltwirtschaft (Edited by R.Lütgens), Vol.III, Ed.2. Stuttgart, German Federal Republic; Frankh'sche Verlagshandlung 332pp.

Otremba, E. (1976) [Commodity production in the world economy.] Die Güterproduktion im Weltwirtschaftsraum. In: Lütgens, R. (*Ed*) Erde und Weltwirtschaft Vol.II and III. Ed.3. Stuttgart, German Federal Republic; Frankh'sche Verlagshandlung 407pp.

Phillips, J. (1959) Agriculture and ecology in Africa. London, UK; Faber and Faber 424pp.

Phillips, J. (1961) The development of agriculture and forestry in the tropics. Patterns, problems, and promise. London, UK; Faber and Faber 212pp.

Piekenbrock, P. (1958) [Vegetation and crop cultivation in the tropics.] Vegetation und Pflanzenbau in den Tropen. *Schriftenreihe der Deutschen Afrika-Gesellschaft* No.7 36pp.

Randall, A. (1975) Growth, resources and environment: some conceptual issues. *American Journal of Agricultural Economics* 57 803–809.

Ruhr-Stickstoff, AG (1953–1957) *Schriftenreihe über Tropische und Subtropische Kulturpflanzen.*

Ruppert, K. (Editor) (1973) [Agricultural geography.] Agrargeographie. Darmstadt, German Federal Republic; Wissenschaftliche Buchgesellschaft 511pp. [Wege der Forschung No.17].

Ruthenberg, H. (1967) [Types of organization of land use and livestock farming in the tropics and sub-tropics.] Organisationsformen der Bodennutzung und Viehhaltung in den Tropen und Subtropen. In: Blanckenburg, P., von; Cremen, H.D. (*Eds*) Handbuch der Landwirtschaft und Ernährung in den Entwicklungsländern Vol.I. Stuttgart, German Federal Republic; Eugen Ulmer 122–208.

Schickele, R. (1931) [Investigations into forms of pasturing in dry regions of the world.] Untersuchungen über die Formen der Weidewirtschaft in den Trockengebieten der Erde. Dissertation, Landwirtschaftliche Hochschule, Berlin 151pp.

Schmithüsen, J. (1961) [General geography of vegetation.] Allgemeine Vegetationsgeographie. In: Obst E. (*Ed*) Lehrbuch der allgemeinen Geographie Vol.IV, Ed.2 revised. Berlin; Walter de Gruyter 262pp.

Spedding, C.R.W. (1975) The biology of agricultural systems. London, UK; Academic Press 261pp.

Tempany, H.; Grist, D.H. (1958) An introduction to tropical agriculture. London, UK; New York, USA; Toronto, Canada; Longmans Green 347pp.

Troll, C. (1952) [Vegetation cover in the tropics and its dependence on climate, soil and man.] Das Pflanzenkleid der Tropen seiner Abhängigkeit von Klima, Boden und Mensch. In: Deutscher Geographentag Frankfurt 1951. Remagen, German Federal Republic; Verlag des Amtes für Landeskunde 35–66.

Troll, C. (1959) [Tropical mountains. Their three-dimensional climatic and geographical vegetation zones.] Die tropischen Gebirge. Ihre dreidimensionale klimatische und pflanzengeographische Zonierung. *Bonner Geographische Abhandlungen* No.25 93pp.

Troll, C.; Paffen, K.H. (1964) [Seasonal climates of the earth.] Karte der Jahreszeiten- klimate der Erde, *Erdkunde* 18, 5–18.

Tsuzuki, T. (1963) [Crop rotation in Japanese arable farming.] Die Fruchtfolgen des japanischen Ackerbaues. *Berichte über Landwirtschaft* 41, 833–846.

Uexküll, H.R. von (1969) [Rice in Asia—problems and possibilities of raising production.] Reis in Asien—Probleme und Möglichkeiten einer Produktionssteigerung. *Zeitschrift für Ausländische Landwirtschaft* 8, 248-259.

Uhlig, H. (1965) [The geographical basis of pasturing in dry tropical and sub-tropical regions.] Die geographischen Grundlagen der Weidewirtschaft in den Trockengebieten der Tropen und Subtropen. *Giessener Beiträge zur Entwicklungsforschung, Reihe 1* No.1 28pp.

Uhlig, H.; Manshard, W.; Gerstenauer, A. (1962) [Contributions to the geography of tropical and sub-tropical developing countries. India, West Africa, Mexico.] Beiträge zur Geographie tropischer und subtropischer Entwicklungsländer. Indien—Westafrika—Mexiko. *Geissener Geographische Schriften* No.2 96pp.

UN, Statistical Office (1970) Statistical Yearbook 1969 New York, USA.

UN, Statistical Office (1977) Statistical Yearbook 1976. New York, USA.

Waibel, L. (1933) [Problems of agricultural geography.] Probleme der Landwirtschaftsgeographie. Breslau, Germany; Verlag Ferdinand Hirt 94pp.

Walter, H. (1939) [Grassland, savannah and the bush in the arid part of Africa and their ecological conditions.] Grasland, Savanne und Busch im ariden Teile Afrikas und ihre ökologische Bedingtheit. *Jahrbücher für Wissenschaftliche Botanik* 87.

Walter, H. (1964) [Tropical and sub-tropical zones.] Die tropischen und Sub-tropischen Zonen. Die Vegetation der Erde in ökologischer Betrachtung. Vol.I, Ed.2. Stuttgart, German Federal Republic; Gustav Fisher Verlad 592pp.

Walter, H.; Lieth, H. (1960, 1964) [Climate chart—world atlas.] Klimadiagramm—Weltatlas. Parts I and II. Jena, German Democratic Republic; VEB Gustav Fisher Verlag.

Walter, H. (1977) [Vegetation zones and climate] Vegetationzonen und klima. Stuttgart German Federal Republic, Verlag Eugen Ulmer 309pp. Ed.3.

Wang, Y.; Nagel, F.; Ruthenberg, H. (1969) [Land use and technical progress in Taiwan.] Bodennutzung und technischer Fortschritt auf Taiwan. *Zeitschrift für Ausländische Landwirtschaft, Sonderheft* No.7 92pp.

Whittlesey, D. (1936) Major agricultural regions of the earth. *Annals of the Association of American Geographers* 26, 199–240.

Wissmann, H. von (1948) [Climatic boundaries of vegetation in the warm tropics.] Pflanzenklimatische Grenzen der Warmen Tropen. *Erdkunde, Archiv für Wissenschaftliche Geographie* 2, 81–92.

Wissmann, H. von (1956) On the role of nature and man in changing the face of the dry belt of Asia. In: Thomas, W.L. (*Ed*) Man's role in changing the face of the earth. Chicago, Illinois, USA; University of Chicago Press 278–303.

# Chapter 4

# Increasing Productivity in Extensive Grassland Farming in semi-arid Areas

## Characteristics of Extensive Grassland Farming

The world acreage under permanent grassland is roughly twice as large as the world acreage under arable cultivation. These grassland areas are mainly used for extensive grassland farming. With this type of management hardy grazing animals feed on the natural vegetation. The farmer does not take any active steps worth mentioning to promote the growth of plants nor does he use buildings to protect his livestock against the inclemencies of the weather nor grow forage nor buy feeding stuffs to compensate for shortages of feed. This chapter deals primarily with market-oriented, extensive grassland farming, and only refers in passing to pastoral nomadism which, although a large category, has little market significance.

Fig. 7
Variations in annual rainfall in Africa. Mean variations as a percentage of the average. (Source: Gregory 1969 p.59)

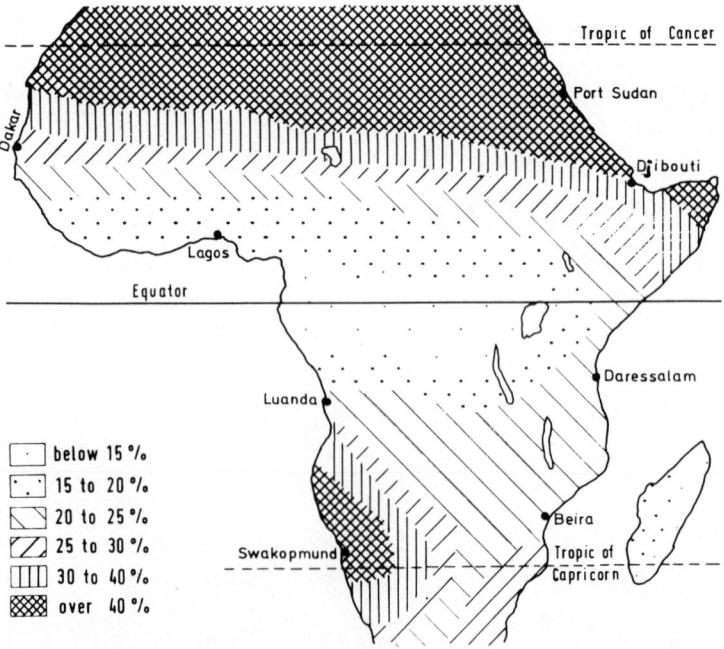

## Geographical distribution

Extensive grassland farming has developed under widely varying conditions of production. It is found in the marginal areas of the tropics where the annual mean temperature is 83° Fahrenheit and in the Kirgiz Steppe, where the monthly minimum ranges from plus 20° to minus 15° Fahrenheit. It is practised by Arabs and Berbers living in climates with monthly temperature minima which are seldom less than 41° Fahrenheit. It is found in the Sahara with its extremely arid climate

Fig. 8
Rain curve for Groot Fontein, S.W. Africa 1899/1900-1962/63; 1 485 m. above sea level, annual average precipitation 534 mm. (Source: von Hase 1964 p.43)

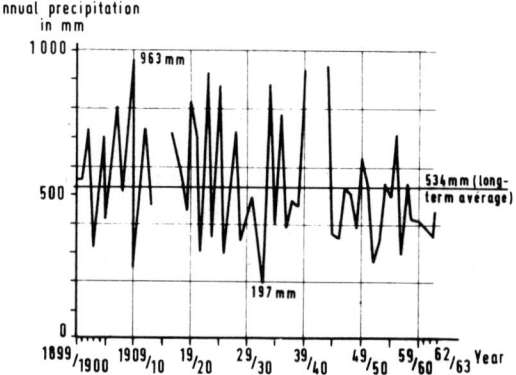

and in New Zealand with its oceanic climate. Agricultural countries such as the Sudan, where the industrialization factor is less than 1, practise extensive grassland farming, as do countries like South Africa with a medium industrialization factor (3.4) and even highly industrialized countries like United States of America where the factor is 9.2. However, extensive grassland farming is most frequently found in the semi-arid zones of the tropics, sub-tropics and moderate climates, i.e. in the dry savannahs and steppes, especially when these areas are situated in developing countries (see Figs 7 and 8).

As regards tropical Africa, the more humid limits of grassland farming are identical with the boundaries between tropical rain forest and sub-humid savannah regions, since cattle and sheep husbandry are hardly possible when there is tsetse fly infestation and when every pasture rapidly reverts to bush and forest. In the more humid areas of the savannah livestock husbandry is more common in places near the climatic dry limit, although such locations are not optimal for livestock husbandry because of the prevailing high grasses and the danger of contagious diseases. Grassland agriculture prevails in the dry savannah, the region between the climatic and the agronomic dry limits; the thornbush steppe is almost exclusively exploited either by livestock farms or by pastoral nomadism and in the semi-desert, even beyond the dry limits of animal husbandry up to the point of transition from semi-desert to desert, pastoral nomadism is met with sporadically.

Table 5   Fodder supply in the African Sahelian zone and its influence on milk yields of Zebu-Azawack cows. (Research results from Toukounous/Niger 1966/67.)

| Characteristics | March | April | June | August | September | November | January | February |
|---|---|---|---|---|---|---|---|---|
| A. Digestible nutrients in plant materials (per kg of dry weight)[1] | | | | | | | | |
| Digestible proteins in g | 4.6 | 4.3 | 4.4 | 117 | 45 | 7.3 | 4.5 | 3.3 |
| starch units (su) | 230 | 240 | 240 | 360 | 280 | 250 | 250 | 250 |
| Digestible protein (su = 1) | 49 | 56 | 53 | 3 | 6 | 35 | 55 | 76 |
| B. Approximate intake allowing for ability to select digestible nutrients[2] | | | | | | | | |
| Digestible proteins | 38 | 20 | 23 | 1760 | 558 | 110 | 27 | · |
| Starch units | 2700 | 2780 | 2600 | 5650 | 3520 | 2800 | 2630 | 2460 |
| C. Protein and energy balance per cow per day | | | | | | | | |
| Digestible protein[3] in g | −450 | −350 | −350 | +1 200 | +50 | −300 | −300 | −300 |
| Starch units | −350 | +200 | +200 | +2 300 | +300 | +200 | +300 | +300 |
| D. Progress of animal's weight in kg | 343 | 316 | 281 | 315 | 336 | 323 | 313 | 304 |
| E. Approximate milk yield in kg per cow per day[4] with calving in: | | | | | | | | |
| Dry season | 3.7 | 3.0 | 2.4 | 4.7 | 3.8 | 2.0 | 1.3 | 1.2 |
| Rainy season | 1.8 | 1.3 | 3.2 | 5.6 | 4.7 | 2.9 | 2.0 | 1.7 |

[1] Rainy season July to September.
[2] Selective grazing produces 30% more protein and energy than the plant composition indicates.
[3] The enormous shortage of protein leads to long periods between calves and irregular calving.
[4] About 50% of the fluctuation in milk yields is due to the inadequate supply of proteins and starch units.
Source: Binzer et al. (1971).

**Productivity**

Commercial, extensive grassland farming is found the world over and has two predominant economic characteristics. (1) It makes extremely high net labour productivity possible. In four different forms of grassland management practised in the United States in the period from 1957/1959 to 1962/1964 labour productivity ranged from 6.10DM to 10.10DM per man hour, with an average of 8.60DM per man hour. Few other farming systems could surpass this. (2) It is possible where the gross productivity of the land is extremely low. An example is Southwest Africa, where land productivity ranged mainly from 1.80DM to 5.50DM per ha in the years 1962/1963. Figures are even lower in many developing countries. It is the farming system with the lowest demand on land productivity.

The reason for the high net labour productivity of extensive grassland farming is the fact that a single worker is able to manage very large areas of grassland with a minimum of material inputs (100 ha of pasture are worked by approximately 0.03 to 0.3 workers). The enormous area of land per worker is possible because in most instances this worker does not do any arable farming but simply tends the natural vegetation. There are no costs for planting, cultivation, etc., and even the costs of harvesting are very small, since harvesting is done by the domestic animal. Work is confined to the supervision and hygiene of the grazing animals. The crop yield per worker is based on very large units of area so that labour productivity is very high even though the crop yields per ha are very low.

The low productivity of the land is of course a negative aspect of extensive grassland farming. A positive aspect is that this kind of agricultural system continues to function in spite of the low productivity of some soils, and makes it possible to make use of areas of low fertility. Another positive aspect is the extremely low minimum intensity of extensive grassland farming, which stems from the low productivity of the soil. This system is thus particularly suited for countries with extremely unfavourable economic conditions (see Fig.9).

**Location of production**

The geographic distribution of extensive grassland farming is determined by two economic characteristics, the possibility of high labour productivity and the toleration of extremely low land productivity. There are two types of area where extensive grassland farming proves superior to all other farming systems—areas where economic conditions are very unfavourable, and areas where natural conditions are very unfavourable. Unfavourable economic conditions of production may be the result of unsatisfactory market prices or of distance from the market. In extensive grassland farming both conditions are usually found.

A farming system of such a low land productivity as extensive grassland farming is typical of thinly populated agricultural countries. Such countries still have sufficient cheap land available so that the costs of using it are low. Workers are scarce and therefore more expensive than in overpopulated agricultural countries. Above all, market prices for all capital goods produced by industry are very high and market prices for agricultural products are relatively low. This price-cost situation is even worse if farm-gate prices are taken into account. Thinly populated agricultural countries have few marketing centres. Railways and roads are scarce and transport rates are high. Since most farms are far away from the market, they are burdened with heavy transport costs.

Fig. 9
Carrying capacity of natural pasture in South Africa. (Source: Cole 1961 p.226)

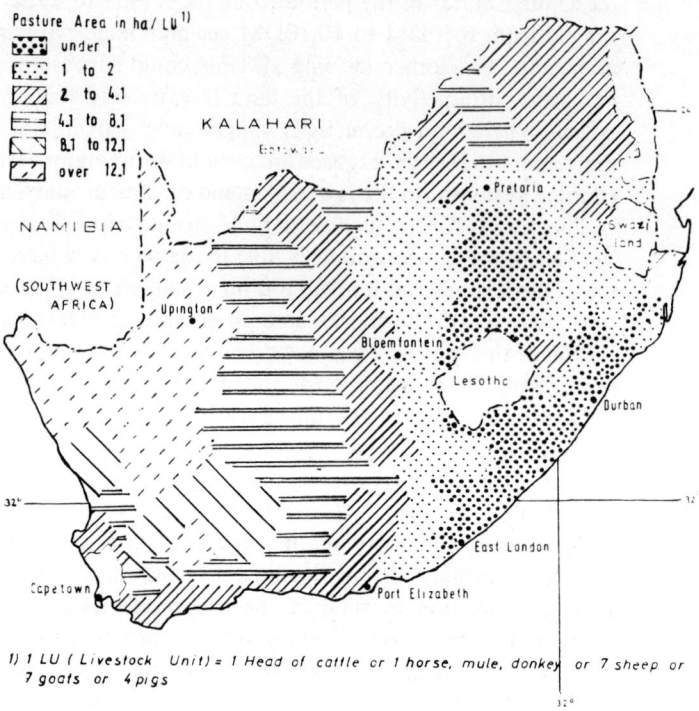

1) 1 LU ( Livestock Unit) = 1 Head of cattle or 1 horse, mule, donkey or 7 sheep or
7 goats or 4 pigs

The farmer is in the dilemma of having to pay both for the transport of his produce to the market and for the transport from the market of the means of production for his farm, which he has purchased there. In other words, he gets less than the market price for his products and has to pay transport costs in addition to the market price for his purchases. The exchange value of agricultural products for manufactured capital goods is therefore very unfavourable. The same is true for labour; it is cheaper near the market.

Typically, therefore, farms far from the market in thinly settled agricultural countries have low costs for land utilization, low prices for their produce, relatively high wages and extremely high prices for buying manufactured capital goods. The principle of marginal productivity therefore suggests they do not need to have a high land productivity, provided they have high productivity of labour and an especially high productivity of capital. The least cost combination can be achieved by farming large areas of land with the minimum of manpower and very sparing use of industrial capital goods. Such a quantitative relationship of factors of production is best achieved by extensive grassland farming. This agricultural system is therefore better than any other system in this type of location.

Extensive grassland farming may also be found where economic conditions are very favourable, for example, in the highly industrialized USA, when the natural conditions of production are so unfavourable that no other farming system is possible or competitive. Extensive grassland farming can also thrive in arid areas where even the grain-fallow system is no longer possible and in semi-arid areas with an annual rainfall of not more than 75 to 100 mm. The lower rate of precipitation constitutes the dry limit of farming. The level of precipitation above which cropping becomes possible lies at 250 to 400 mm. in Africa. Where the distribution of rainfall is uneven, this critical level is much higher. In parts of east Africa it lies at 600 mm.

**Types of production**

Fig. 10 illustrates how the type of production in extensive grassland farming is determined primarily by the distance from the market.

The straight lines falling off to the right show how the level of profitability sinks as the distances to the market increases. The less transportable the products the

Fig. 10
Profitability of types of extensive grassland production in relation to distance from the market

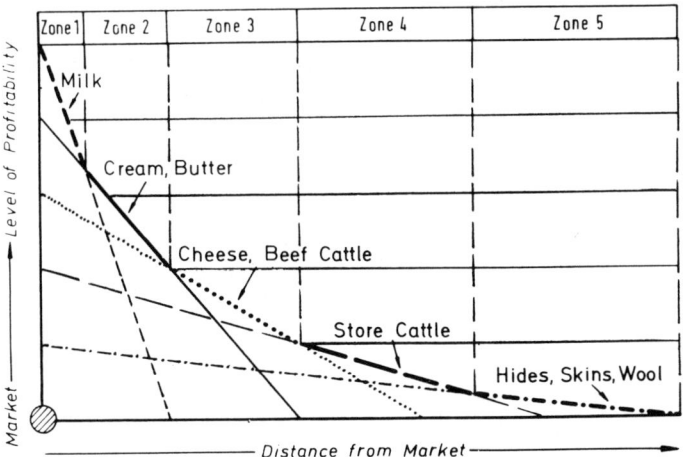

more rapidly the level of profitability sinks. On the other hand, the less transportable a product is the more profitable it is in the immediate vicinity of the market. High prices must be paid for such products so as to make sure that demand will be met.

In Zone 1, nearest the market, milk production is the most profitable. In Zone 2, the farmgate value of milk has fallen to the point where it is more advisable to skim the milk and concentrate on the production of butter and cream. Neither whole milk nor skimmed milk can be marketed profitably from this distance. In Zone 3 both butter and cream production have lost their high level of profitability and should be restricted in favour of cheese and beef cattle production. Store cattle production can also be considered in this zone, since transport costs can be eliminated and the cattle be taken to market on the hoof. In Zone 4 only store cattle production is advisable from the point of view of profitability as labour and capital requirements are lower than in all the other types of production considered. In Zone 5 only those types of production are profitable which yield easily transported products such as hides, skins and wool.

This shift in the competitiveness of sheep and cattle husbandry products quite naturally determines the type of production on the farms in the various zones. Of course it cannot be argued that only one type of production is feasible in each zone. It is clear, however, that there is very little scope for product diversification in extensive grassland farming, particularly when it is realized that rainfall is also a factor which tends to restrict diversification.

The various types of production in sheep and cattle husbandry make very different demands with respect to the quality and quantity of forage. Each type of production is best suited to specific types of pasture and these, in turn, are determined by the level of rainfall available. In the case of Southwest Africa where the annual rainfall is only 100 to 250 mm and the scarce vegetation is suitable only for sheep husbandry, sheep farming is to be preferred to cattle farming not only because sheep are less exacting animals which can cover long distances in search of feed but also because their pointed muzzles enable them to select the more nutritive parts of the forage plants. In addition, they can manage with less concentrated nutrients because of the nature of their digestive tract. The sheep are raised for wool and skins.

When the annual rainfall rises to 250 to 300 mm cattle husbandry is now possible and production is concentrated on store cattle. Calf production is not yet feasible because the cow is not capable of calving more than once every second or third year at the most, because of the scarcity of fodder. Calf production cannot be profitable with such a calving rate of 33 to 50%. Calves are, therefore, usually bought from other regions with richer vegetation. The specialized farm rearing store cattle is found under these conditions.

In rainfall zones with 350 to 400 mm annual rainfall the self-replenishing beef cattle farm is typical. The quality of the pasture is so much better that fattening and calf production are both possible. To compensate for risks and to get around the difficulties involved in buying calves from farms in other climatic zones the farms become more diversified. The farmer maintains his own breeding herd, rears all the calves and markets them as half-finished fatstock.

In the climatic zone where annual rainfall reaches 450 to 500 mm conditions for rearing calves are so favourable that if the prices are sufficiently attractive they can profitably be produced for the market. These farms supply the store cattle farms in the dry zones (250 to 300 mm) with calves. The farmer may even change over to calf production entirely, selling his calves at the age of 9 to 18 months to specialized fattening farms.

With an annual rainfall of 500 mm and above cropping becomes possible. This means that forage can be grown on arable land, and the farmer can improve the nutritive value of the fodder and establish feed reserves for the dry seasons. The farms in this climatic zone are dairy farms and high grade beef cattle farms. The dairy farms must of course have ready access to market outlets for milk or for butter and cream.

## Increasing Production with Extensive Grassland Farming

### The most effective means of alleviating feed shortages

The typical characteristics of extensive grassland farming are absence of crop production, the restriction of husbandry to only one type of livestock—sheep, goats

Table 6    Development of grassland farming in Southwest Africa in relation to annual rainfall.

| Characteristics | Unit | A.  pastoral sheep farming | | | B.  pastoral cattle farming | | |
|---|---|---|---|---|---|---|---|
| **General features** | | | | | | | |
| Natural zone | | Kalahari-Thornbush savannah | | | Damaraland-Thornbush savannah | | |
| Position of farm | | Southeast Windhoek | | | South Seeis | | |
| Farm size | ha farmland | 10 000, from 1964/65 6 000 | | | 11 500 | | |
| of which arable | ha farmland | – | | | – | | |
| Ranking of farming enterprise | | 1.  production of karakul pelts<br>2.  breeding rams<br>3.  beef cattle production with nurse cows | | | 1.  beef cattle production with nurse cows<br>2.  some sale of milk | | |
| Rainfall average over several years | mm/year | 220 | | | 350 | | |
| Crop year | | 1962/63 | 1964/65 | 1965/66 | 1962/63 | 1964/65 | 1965/66 |
| General description | | Consecutive year of a drought period | Good average year | Very good year | Consecutive year of a drought period | Good average year | Very good year |
| Rainfall | mm/year | . | . | 250 | . | 306 | 352 |
| Rainfall previous year | mm/year | 50 | . | . | 209 | . | 306 |
| Water holes | number | 10 | 10 | 12 | 3 | 3 | 5 |
| Camps | number | 16 | 16 | 47 | 8 | 8 | 10 |
| **Farm organization** | | | | | | | |
| Livestock numbers | LU | 930 | 908 | . | 1 660 | 1 323 | . |
| Area of pasture | ha/LU | 11 | 7 | 6 | 7 | 9 | 8 |
| Pasture utilization | | | | | | | |
| by cattle | % | 22 | 2 | 12 | 90 | 91 | 96 |
| by sheep | % | 74 | 96 | 87 | – | – | – |
| by other livestock | % | 4 | 2 | 1 | 10 | 9 | 4 |
| Calving results | % | 38 | 88 | 87 | 53 | 71 | 83 |
| Cattle sales (net) | % of herd | 6 | . | . | 16 | . | . |
| Lambing results | % | 98 | 147 | 88 | – | – | – |
| Karakul pelts sold | % of flock | 30 | 56 | 53 | – | – | – |
| **Farm results** | | | | | | | |
| Gross returns | DM/LU | 143 | 215 | 209 | 39 | 53 | 77 |
| Input costs | DM/LU | 39 | 64 | 70 | 10 | 14 | 17 |
| Wages for outside Labour | DM/LU | 15 | 21 | 19 | 7 | 8 | 9 |
| Farmer's gross income | DM/LU | 89 | 130 | 120 | 22 | 31 | 51 |

Conversion key for Livestock units (LU); head of cattle = 1; sheep = 0.16;
goat = 0.20; donkey = 0.50; horse or mule = 1.20; pig = 0.50

Source: Enquiry by M. Buerger, Business and tax advisor in Windhoek, Southwest Africa.

or cattle—and the fact that the particular type of animal is usually kept for only one production purpose. This is clearly a system of mono-production with all its related disadvantages and dangers. Production is not diversified nor does the production process encompass many stages and the market risk is therefore correspondingly great.

In the semi-arid areas the frequently extreme fluctuations in rainfall from year to year also burden extensive grassland farming with a high production risk. This is particularly so in those regions where the conservation of fodder or the purchase of concentrates are not viable activities. Here the farms are completely dependent on the natural vegetation. It becomes obvious how acute this problem is when we realize that in the main geographical regions for extensive grassland farming, the dry savannahs and steppes, a very short wet season is followed by a very long dry season even in years with a normal rate of precipitation. The main problem facing every farmer is how to bring his livestock through the long dry season with as few losses as possible. Extensive grassland farming is thus jeopardized both by heavy production risks and heavy market risks.

The problem of market risk has to be tackled primarily within the framework of general market policy but the problem of production risk must be approached as one of farm management. To reduce the production risk the main problem is to carry the livestock population over the dry season with a minimum of economic losses. A successful solution is the surest way to increase productivity.

Positive action to overcome feed shortages can either involve taking counter-measures or making passive adaptation. In industrialized countries feed shortages are usually overcome by countermeasures. Root crops and other forage crops are grown and stored for the season of feed shortage, some being preserved in the form of hay and silage. Alternatively fodder is purchased. The price-cost relationship involved must naturally be favourable.

Passive adaptation has to be made to feed shortages if the price-cost ratios, are unfavourable, i.e., if the purchasing power of the farm produce is so low that neither forage conservation nor fodder purchases are economically feasible. The farmer accepts the limits set him by natural conditions and selects that type of grassland husbandry which best enables him to carry over his livestock. This method of passive adaptation is necessary when grassland farming is especially extensive, as is largely the case in developing countries. Passive adaptation to alternating seasons of feed shortage and abundant feed involves the following five groups of measures:

(1) That species of animal is selected which possesses a great capability to endure and overcome periods of feed shortage. Thus the desert nomads keep camels and sheep which can go without food and water for long periods of time. In Ethiopia we find many goats because they can feed on bush foliage in periods of drought.

(2) Hardy breeds of animal are given preference. Fat-tail sheep and zebus, for example, can endure long lean periods because they are able to develop fat reserves in times when feed is plentiful. In other words, they stockpile their own nutrients.

(3) Types of production are chosen which are less dependent on the seasonal food balance. For example, the farmer concentrates on store cattle rather than on dairy production. If he is raising sheep, he concentrates on the production of wool and skins rather than meat.

(4) Animal performances are limited to seasons when feed is abundant. The

calving intervals, for example, are so determined that calving takes place at the beginning of the wet season. This ensures that the dams do not suffer from hunger during the strenuous suckling period. The final stage of fattening takes place in the wet season.

(5) Livestock density is regulated according to seasonal fluctuations of forage production. Livestock densities are increased at the beginning of the rainy season and decreased during the dry season. Provisions are made for both births and the purchases of animals to take place at the beginning of the wet season while slaughter and sale of animals are carried out at the beginning of the dry season.

In addition to these five methods of passive adaptation to feed shortages, other farm management methods which vary according to climatic zones can be used to achieve a feed balance.

## Feed balance in climates where arable farming is possible

As one example there are areas in Argentina where the annual rainfall is sufficient to make the cultivation of land possible and where only the great distance from the market makes it unprofitable to grow crops. In the past owners of large haciendas in these areas leased out small areas of grassland to subsistence farmers on the condition that they were sown with alfalfa and returned to the landowner after three or four years of cropping. In this way the estanciero obtained alfalfa fields for the final stage of fattening his cattle during the drought period. Alfalfa with its long tap-roots which can utilize groundwater better than the roots of ordinary grasses was well suited for this purpose. The settler in his turn profited from the arrangement because he benefited from the high yield of the soil and did not have to bear the costs of soil regeneration.

Another example may be taken from Colombia where this symbiosis between large and small holdings still exists. In the coastal lowlands the annual precipitation of not more than 400 mm is distributed over only four months. Natural vegetation consists of thorn-bush savannah with succulents and cactus shrubs, with occasional thorn-trees. Haciendas in this area range from 1000 to 50 000 ha in size and their grasslands are exposed to the danger of reverting back to bush. For this reason, the owners turn over part of their land to small leaseholders for a period of 18 months. The tenants clear and burn the bushy grassland and cultivate it with maize, cassava ("yucca"), fibre plants etc. In exchange for the right to use the land, the leaseholder sows the land with guinea grass and returns it to the hacienda owner as an excellent pasture after a period of 18 months.

Cooperation between large and small holdings is necessary because of the differences in the scarcity ratio of factors of production in the two categories of farm. The estancia owner has large tracts of land but a scarcity of labour. It is in his economic interest to substitute land for labour by renouncing the labour-intensive land clearing operation at the cost of land utilization. On the other hand, the small leaseholder is short of capital and does not own any land but he has sufficient family labour at his disposal. It is to his economic advantage not to have to pay in cash for the utilization of the soil but to exchange the privilege of land use for the labour of clearing it and sowing grass on it.

A third similar example came to my notice on my visit to Rhodesia in 1965. An experienced grassland farmer generously allotted large areas of land to his farm-hands for growing maize but shifted these plots of land on the pasture ground every year. The relatively small loss in forage in the wet season did not worry him since at that time of year he had plenty of forage for his animals but during the dry season he was very thankful to have some maize stubble pasture. This was in itself of no use to the part-time crop farmers as they did not possess any livestock for grazing. In addition to this the bent-grass (eragrostis curvula) which grows especially well on soil which has been temporarily used for crop farming is particularly rich in nutritive substances in the dry state and hence is very good for winter pasture.

It is clear that the above three examples have the same basic principle in common. Grassland conditions in the dry season can be improved by temporarily introducing arable farming with subsequent reseeding of grass or appearance of natural grass flora. Because of the unfavourable market situation, the owners of large ranches cannot profitably grow crops. They therefore leave tillage to small subsistence farmers who consume the farm products themselves and are not dependent on the state of the market.

## Feed balance in climates where arable farming is not possible

In dry areas which are not suitable for arable farming and where there is no great differentiation in farm sizes, these means of overcoming feed shortages cannot be used.

Fig. 11 shows nine stages of extensive grassland farming each of which is typical of a successive stage in the growth of the national economy. Eight different techniques used to overcome feed shortages in dry seasons are shown, classified by the increasing rate of land productivity.

### Feed balance by means of land reserves

As long as labour and capital are very scarce and land is abundantly available, feed must be balanced by means of land reserves. Stage I is that of pastoral nomadism. The population has not yet exchanged a nomadic mode of life for a settled one. There is very little ownership of land. This traditional way of life is characteristic of extremely arid areas where transport conditions are very unfavourable. Typical for this group are the Bedouins roving in the steppes of Syria, the Kurds and the Berber tribes. In East Africa, the Masai, and the Somali and in West Africa the Fulbe can at least be classified as semi-nomads. Nomadism is caused by the necessity of achieving a feed balance over the course of the year through driving the herds across the country. Nomadism is practised particularly where the rhythm of annual vegetation varies from region to region, because of the presence of mountain chains, highlands and lowlands. In general the movements of montane nomads are vertical, those of lowland nomads (in the Sahara, e.g.) are horizontal. At this stage domestic animals, like the natural game species, must wander long distances to find feed throughout the year. None of the nomad's domestic animals, usually camels, fattail sheep, goats, or donkeys, are exacting as to their requirements. He lives off them and accompanies his herds on their wanderings.

When at later date (Stage II), the nomads take up a settled way of life, a system

Fig. 11
Stages of fodder balance in grassland farming in semi-arid climates. (Source: Andrea 1965, p.92)

## A. Methods of Passive Adaptation to Seasons of Fodder Shortage

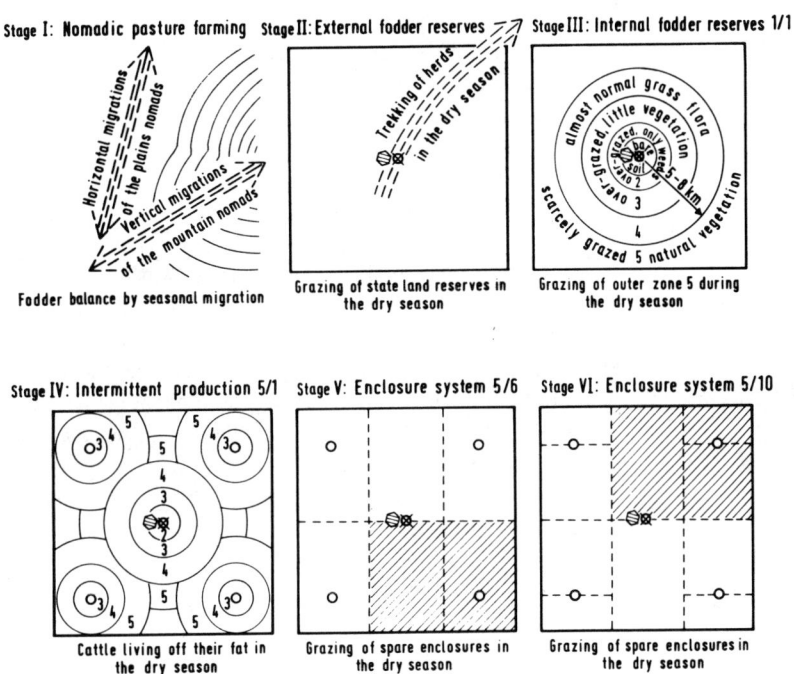

## B. Methods of Actively Overcoming Times of Fodder Shortage

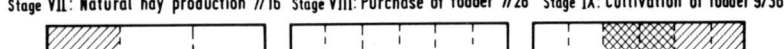

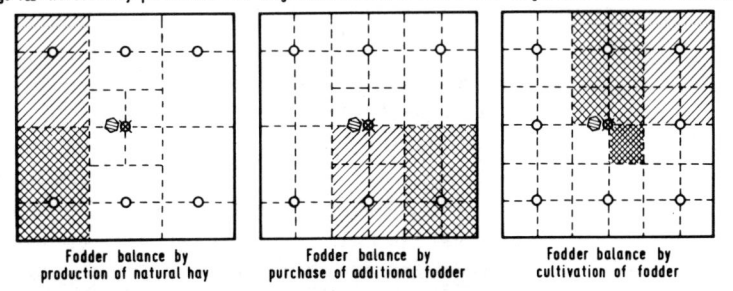

of land ownership develops. Plots of land are measured out and allocated to individual farmers. But as long as the area is not densely populated, not all the land is in private hands. The state still retains land reserves which can be opened up as pasture land when food is scarce during a drought.

At Stage III the situation again changes. All areas have now been turned over to private ownership. Farms are fenced off although pasture land is not yet subdivided. If the farm in Stage III has only one watering place, then different zones of vegetation develop around this centre. The soil close to the watering place, the first zone, is barren, since the concentration of dung and urine and the damage caused by treading puts an end to any growth of plants. Even the second zone has little vegetation as the soil mostly harbours weeds of little value. The third zone, although quite far away from the watering point, is still overgrazed and has many weeds. However, vegetation is growing denser, and there are already some good grasses. Nearly normal vegetation with few weeds and many good grasses is found in the fourth zone. Animals reaching this zone have as a rule arrived at the outer boundary of an area that can be grazed within reach of the watering place. The fifth zone is not used during the wet season when feed is plentiful, the distance to the watering place being too great. A cover of natural vegetation therefore flourishes. During the dry season when grass is scarce, this zone acts as a feed reserve. This is undoubtedly true in the outer belt of the tropics where the dry season coincides with winter and where, since the animals do not need as much water in the cool season, the grazing radius can be larger in winter. Another reason why the grazing radius increases is that by the beginning of the dry season the calves and lambs have become fit to cover longer distances.

*Feed balance by means of capital input*

In the course of economic development land tends to become scarcer and more costly. This calls for increased utilization of the productive capacity of the soil and this also becomes possible by using the capital inputs which have also meanwhile become more plentiful. At Stage IV wells are sunk to supplement existing watering places and this steps up animal production in the following three ways:

(1) The distance to be covered by the cattle from the grazing grounds to the watering places is shortened. The energy which is no longer needed for travelling is channelled into improved market performances, i.e., fattening and wool growing.

(2) Zone 1 of Stage III which was barren of vegetation disappears completely. At less frequented subsidiary watering points Zone 2 which was almost valueless for feeding purposes also disappears since each of the watering places is less heavily frequented by the animals.

(3) Zone 5, which was seldom utilized also disappears since all areas may now be grazed, even during the wet season.

The farm's vegetation cover thus becomes more even and, in the wet season, richer so that it is possible to increase the number of livestock. On the other hand in the dry season feed supply conditions have grown worse since feed reserves are no longer maintained either outside or inside the farm.

At this stage animals live on starvation diets during the dry season. Under prevailing price-cost ratios this has simply to be accepted as necessary. However, land is still cheap compared with labour and capital so that production must be as labour and capital-extensive as possible, even when this requires large tracts of land. This goal is best achieved by means of intermittent production. During the dry season not only is no meat produced, but loss of weight, which must of course be regained when the wet period sets in, is also accepted as necessary.

Intermittent production has the disadvantage that the animals do not reach their slaughter weight until the age of four or five years. As a result there is a disproportionate increase in the total amount of feed required to maintain the animals and the productive feed intake is relatively small. Large tracts of grassland per 100 kg increase in live-weight are, therefore, still required. Nevertheless, this method of

Fig. 12
Development in the weight of beef cattle in Namibian ranching

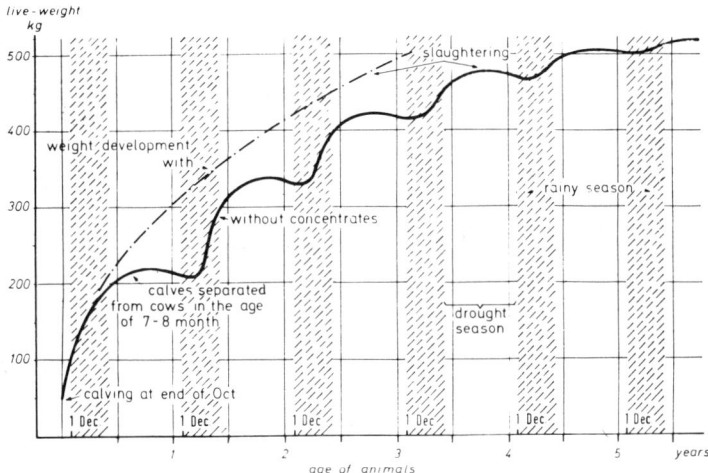

grassland management is economically sound; it is the only system which uses the minimum of labour and capital per 100 kg of slaughter animal i.e., it is the minimum cost system.

Only when land becomes more expensive and capital cheaper does it pay to make larger capital investments which ensure better utilization of the soil.

Stages V and VI have now been reached, and it becomes practicable to enclose the grassland area by fences. The investment required for the installation of fences is considerable and sheep must be protected by jackal-proof fences which are especially expensive. The use of the camp technique, however, makes possible controlled grassland management and hence fuller utilization of the land. It now becomes feasible to provide pasture of varying qualities for the various species of animal, types of utilization, and age groups according to their feed requirements. Reserve camps can be established and are rotated every year and not used for grazing during the wet period. In the dry season these camps are opened to the animals in times when feed is scarce.

When this method of overcoming feed shortages is used, the animals no longer lose weight during the dry season and in fact, frequently register an increase in live

Table 7   Stages of intensity in Southwest Africa extensive grassland farming classified by number of "camps". Crop year 1965/66. (Predominant type of farm: self regenerating beef cattle farms with subsidiary karakul sheep.)

| Farm | Farm size to the nearest 500 ha | Annual precipitation in mm (annual average) | Number of watering places | Number of camps | Average area of camps in ha | Area of grassland in ha/LU | Calving results in % | Lambing results in % | Gross returns in Rand/LU |
|---|---|---|---|---|---|---|---|---|---|
| A | 6 000 | 320 | 9 | 3 | 2 000 | 9 | 54 | – | 9.23 |
| B | 11 500 | 350 | 5 | 10 | 1 150 | 8 | 83 | – | 13.95 |
| C[3] | 8 500 | 420 | 4 | 11 | 773 | 9 | 81 | – | 15.83 |
| D[3] | 11 000 | 395 | 8 | 12 | 917 | 8 | 66 | – | 14.18 |
| E[4] | 5 500 | 420 | 5 | 12 | 458 | 6 | 43 | 121 | 14.35 |
| F | 12 500 | 350 | 10 | 13 | 962 | 7 | 70 | – | 11.72 |
| G[4] | 9 000 | 350 | 7 | 13 | 692 | 12 | 95 | 98 | 14.69 |
| H[6] | 16 300 | 163 | 6 | 17 | 959 | 25 | 88 | 132 | 13.41 |
| I | 7 000 | 280 | 6 | 18 | 389 | 10 | 44 | 119 | 20.14[2] |
| J[4] | 12 000 | 250 | 4 | 19 | 632 | 10 | 69 | 98 | 8.31 |
| K[6] | 18 000 | 125 | 10 | 20 | 900 | 18 | 57 | 96 | 16.92 |
| L | 11 000 | 220 | 6 | 22 | 500 | 14 | 96 | 120 | 10.76 |
| M[4] | 17 500 | 320 | 6 | 24 | 729 | 13 | 71 | 123 | 10.37 |
| N | 5 000 | 450 | 9 | 24 | 208 | 8 | 68 | – | 15.45 |
| O[5] | 10 500 | 500 | 7 | 25 | 420 | 11 | 69 | – | 20.84 |
| P | 18 000 | 350 | 11 | 26 | 692 | 9 | · | 95 | 9.24 |
| Q | 13 000 | 320 | 7 | 29 | 448 | 9 | 40 | 122 | 4.98[2] |
| R | 10 000 | 200 | 9 | 32 | 313 | 14 | 64 | 107 | 8.50[2] |
| S | 18 500 | 250 | 9 | 36 | 514 | 11 | 68 | 106 | 14.43 |
| T | 15 500 | 420 | 11 | 39 | 397 | 8 | 82 | – | 13.95 |
| U[6] | 6 000 | 220 | 12 | 47 | 128 | 6 | 87 | 88 | – |
| V[6] | 10 000 | 220 | 18 | 63 | 159 | 12 | 59 | 121 | 53.50[2] |
| Average | 11 468 | 313 | 8 | 23 | 652 | 11 | 69 | 110[7] | 14.99 |

[1] LU = livestock unit; bullock = 1; sheep = 0.16; goat = 0.20; donkey = 0.50; horse and mule = 1.20
[2] Figures distorted due to exceptional sales of livestock etc. and purchases related to undervaluation for taxation
[3] Speculative purchases of livestock
[4] Karakul sheep on less than 20% of the total grassland
[5] Intensive milk production with cultivation of additional fodder for cattle and pigs
[6] Sheep breeding enterprise with significant sales of rams (karakul)
[7] Average of the sheep farms

Source: survey by M. Buerger, Business and Tax advisor in Windhoek.

weight, so that cattle reach slaughter weight in three years. This makes it possible to make better use of the land than in Stage IV. The proportion of feed used for maintenance in the production of three-year old cattle is much less than for the production of five-year olds of the same weight class, so that the proportion of fodder which is productive increases. The increased productivity of grassland which is achieved at this stage can also be attributed to the fact that the soil remains more fertile than at Stage IV, so that more forage is grown per ha.

*Feed balance by means of labour input*

Stage VII of Fig. 11 differs from Stage VI in two ways. Firstly, the number of watering places has been increased in the course of intensification from five to seven, and the number of camps from 10 to 16. Secondly, a limited amount of hay is now produced from natural grass. A more advanced stage of economic develop-

Fig. 13
Devonby Farm, near Gobabis, Namibia, 1970 (6139 ha)

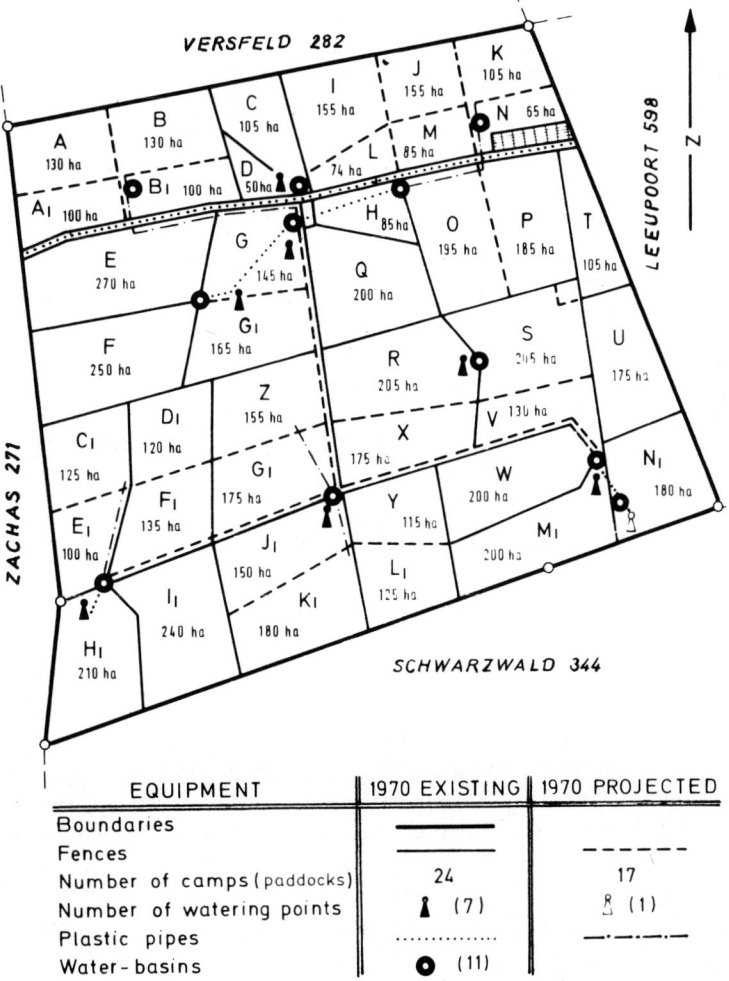

| EQUIPMENT | 1970 EXISTING | 1970 PROJECTED |
|---|---|---|
| Boundaries | —— | |
| Fences | —— | - - - - - |
| Number of camps (paddocks) | 24 | 17 |
| Number of watering points | ⚑ (7) | ⚑ (1) |
| Plastic pipes | ·············· | —·—·—·— |
| Water-basins | ● (11) | |

ment can now be reflected not only by a further increase in the amount of capital invested but also by an increase in labour input.

Part of the natural grass growing on areas where grazing is not permitted during the wet season is now mown and the hay is stored in small kraals for use as fodder in the dry season. This hay is richer in nutritive value than standing hay, as mowing is done at a time when the nutrients have not yet moved down to root stocks or seeds but are still in the stalks and leaves.

In the process of economic development, the production of hay from natural grass is the first active measure taken to overcome feed shortages. It is possible only when the purchasing power of animal products in terms of labour has also increased as it involves high labour costs as well as high costs for machinery and traction in semi-arid zones where vegetation is scattered.

By Stage VIII the number of camps has been increased to 28, and the camp technique has been further developed and refined. At this stage the reserve camps used to overcome feed shortages in dry seasons are supplemented not only by hay produced from natural grass but also by purchased feeds. At this stage of economic development the second active measure to overcome feed shortages becomes economically feasible, namely, the purchase of feed for the dry season.

Population growth and the process of industrialization increasingly stimulate the growth of towns. Near industrial centres and human settlements intensive irrigated farming is practised in order to supply the urban population with basic foods. For reasons of crop rotation some lucerne is cultivated under irrigation and from six to eight crops a year are cut. As soon as the transport network has developed sufficiently bales of compressed lucerne hay are sold to extensive grassland farms.

At the next level of industrialization the costs of land have increased still further so that it is even more important for extensive grassland farmers to step up the productivity of land. At this stage, however, the exchange value of slaughter animals in terms of concentrate feeds has become so favourable that the farmer is in a position to purchase concentrates as well as hay. These concentrates will probably consist of grain-maize brought in from regions where arable farming is possible without irrigation.

Since the national transport infrastructure has also developed further in the wake of increasing industrialization, the transport of grain-maize which is a fairly transportable commodity is now economically feasible. The feeding of animals with concentrates has now become profitable for another reason. Increasing industrialization goes hand in hand with a rise in mass income, and the resultant increase in meat consumption brings with it better prices for high-quality slaughter animals. Extensive grassland farmers who are normally unable to produce slaughter animals of high quality because of the low nutritive value of the forage growth, now find it worthwhile to supply their animals with concentrates in the dry season and thus to produce better quality meat in the final stage of fattening.

At Stage IX the number of camps has now been raised to 36. This measure is particularly expensive, since it involved the sinking of two additional wells. In the humid savannah there are very often ponds and brooks providing water for the herds. In the dry savannah, however, even larger rivers dry up making it necessary to drill for groundwater. This water is frequently pumped up using windmills.

The principal difference between Stage IX and Stage VIII is that in Stage IX a third method is used to overcome feed shortages. Forage is grown with irrigation near the farm buildings. The main well, which is no longer fully required for drinking purposes is now also used to supply irrigation water. Growing forage as a field crop is very labour intensive and presupposes an advanced stage of overall economic development. Thus it requires a high purchasing power for animal products in terms of farm labour as well as in terms of farm supplies like machinery and mineral fertilizers.

When the distance of the farms from the market is very great, forage production (Stage IX) must precede the purchase of feed (Stage VIII). Stage IX of Fig. 11 shows the final stage of extensive grassland farming which forms a bridge to the intensive forms of grassland farming.

Table 8 shows the farm management arrangements open to the farmer in successive stages of general economic development which enable him to overcome seasonal shortages of feed and increase the productivity of the land in grassland farming in semi-arid climates.

## Increased Productivity through Eliminating Extensive Grassland Farming

Finally we must examine the conditions under which extensive grassland farming is replaced by more productive types of farming, look at the types of farming concerned and consider how and to what extent a country's development policy can influence this evolution.

### Motive forces of the evolution

The motive forces which bring about agricultural evolution and the generating forces of economic life in general are either changes in the price-cost ratios or technological progress. As economic development gradually advances these two motive forces also increasingly reduce the economic feasibility of extensive grassland farming. In locations where natural conditions of production are also suitable for other types of farming extensive grassland farming can flourish only when the purchasing power of agricultural products in terms of labour and agricultural inputs of industrial origin is extremely unfavourable. An increase in purchasing power which provides the basis for a changeover to more productive types of farming can result from one of two changes.

Market prices for agricultural products can rise because of population growth or because industrialization has increased the purchasing power of a stable population. Such a rise in market prices automatically affects farmgate values regardless of the distances from the market and this, in turn, means that the relative costs for farm labour and agricultural means of production provided by industry are lower.

Alternatively the development of transport facilities and higher densities of settlement of the economic area reduce marketing costs and bring the farm, so to speak, nearer the market so that the price-cost ratios improve noticeably in favour of the farmer. Every new railway line, every new road, every reduction in transport rates automatically reduces the differences in transport costs between farms at different distances from the market. Farmgate values for agricultural products rise and the agricultural means of production purchased by the farmer at the market cost less to be brought to the farm.

Table 8   Management arrangements to overcome seasonal feed shortages in extensive grassland farming in the course of economic development.

| Stage of development of the national economy | Stages Fig. 11 | Price − = low prices + = high prices | | | Quantity − = low input + = high input | | | Passive main measures to overcome feed shortages | Active main measures to overcome feed shortages |
|---|---|---|---|---|---|---|---|---|---|
| | | Land | Labour | Capital | Land | Labour | Capital | | |
| Thinly populated agricultural countries | I to III | −− | −− | ++ | ++ | + | −− | Hardy types and breeds of animals (zebus, fat-tail sheep, etc.); selection of calving periods; intermittent production; seasonal adjustment of livestock density. Pastoral nomadism; transhumance; farmfeed reserves | none |
| Beginning of industrialization | IV to VI | − | − | + | + | −− | − | Hardy types and breeds of animals; selection of calving periods; fences and watering points; reserve camps | none |
| Industrialized agricultural countries | VII to VIII | + | − | − | − | + | + | Selection of calving periods; additional camps and watering points; additional reserve camps | Production of hay from natural grass; purchase of alfalfa hay; later on, purchase of grain feed as well |
| Industrialized countries | IX | ++ | ++ | −− | −− | + | ++ | More additional camps and watering points; improved camp technique, in particular reserve camp technique | Field forage growing on small irrigated areas; later on also growing of millet, Bechuana beans, etc., and where possible, maize |

When the purchasing power of agricultural products has reached a certain level and when the possibilities of extensive grassland farming prove too small to take full advantage of the opportunities of intensive management opening up, extensive grassland farming is replaced by other types of farming that are more productive. Technological progress also leads to a reduction in the land area available for extensive grassland farming. Water development schemes in the dry regions of Africa and technical progress in the field of irrigation have increased the competitiveness of arable farming. The mechanization of draught power and the introduction of harvesting machinery have also made it possible for arable farming to become competitive and gradually to take over large areas of land which were originally used for extensive grassland farming.

**Nature of the evolution**

The way extensive grassland farming evolves varies in nature according to the climatic zone. Where there is sufficient rainfall to allow arable cultivation, millet and grain-maize crops are adopted in extensive grassland farms. This at first occurs in the form of a monoculture, since millet and grain-maize are self-tolerant. Monocultures of millet and of grain-maize on an extensive grassland farm, encouraged both by technological innovations which make tillage cropping more advantageous and by better price-cost ratios demanding intensification, gradually reduce the natural grassland area of that farm. Millet and grain-maize cultivation eventually reaches the more hilly areas and penetrates into areas where the soil is less fertile, until at last the point is reached where it no longer pays to plant them in a one-crop system. Yields are so low and costs are so high that it is more sensible to change over to crop rotations.

In order to achieve this, grass and legumes are introduced into the rotations, at first with the primary object of improving the biological condition of the soil. The provision of fodder is of secondary interest. Ley farming is intended primarily to improve the soil for grain crops rather than to help cover feed shortages in dry seasons. It is clear, however, that ley farming leads to grazing grounds being made available in the dry seasons and hay being produced. This then is the first step towards uniting two previously separate types of production, beef cattle and grain. Leys not only make rotation farming possible but also improve the dry-season fodder base for fattening cattle. Soil under ley becomes enriched with root humus which is beneficial to subsequent tillage crops. Furthermore, the excreta of livestock grazing on the leys also improve the nutrient content of the soil intended for grain production. Two types of enterprise, animal husbandry and crop husbandry, are thus effectively associated and combined into a single system.

At a later stage of development, field crops which are more productive such as soybeans, groundnuts, sunflowers, etc., are grown. Arable farming spreads until finally all the natural grassland area which is at all suitable has been converted to arable farming. Determinant factors are degree of incline, depth of surface soil, groundwater table etc.

In the dry savannahs, the dry steppes and the semi-desert regions the lack of adequate rainfall makes arable farming impossible. Consequently, tillage cropping cannot threaten extensive grassland husbandry. In such hot and dry zones where no millet and grain-maize monocultures can thrive, the evolution from extensive grassland farming to arable cultivation is brought about in a single leap by introducing irrigated farming.

## Possibilities and limits of development policy

From what has been said it is not difficult to recognize the possibilities open to development planners and to realize where the limits are. In order to progress from extensive grassland farming to more productive farming, it is necessary that price incentives be created for the farmer. Since he is concerned with farmgate prices and not with market prices, the price-cost ratios can be improved not only by raising the level of market prices for agricultural products and pushing down the prices for agricultural means of production but also by developing and expanding the transport network. In other words, as experience has shown, the best way to improve the price-cost ratio in favour of the farmer is in the first instance to implement a sound industrial policy and in the second instance to implement an effective infrastructure policy at the same time.

In many countries where extensive grassland farming is predominant such a development policy is obviously required. Extensive grassland farming does have the advantage of high labour productivity but against this is the disadvantage of very low productivity of the soil. It can provide work and food for very few persons per square kilometre. Therefore in an epoch of population explosion such as ours efforts must be made in large parts of the dry savannahs and steppes and even the semi-desert regions to progress beyond extensive grassland farming to more productive types of farming.

## Bibliography

Andreae, B. (1965) [Land fertility in the tropics. Utilization and maintenance. Farm management considerations for work in developing countries.] Die Bodenfruchtbarkeit in den Tropen. Nutzbarmachung und Erhaltung. Betriebswirtschaftliche Überlegungen für die Arbeit in Entwicklungsländern. Hamburg, German Federal Republic; Paul Parey 124pp.

Andreae, B. (1966) [Pasturing in southern Africa. Studies of location and development theory of the agricultural geography of the tropics and sub-tropics.] Weidewirtschaft im südlichen Afrika. Standorts- und evolutionstheoretische Studien zur Agrargeographie der Tropen und Subtropen. *Geographische Zeitschrift, Beihefte Erdkundliches Wissen* No.15 50pp.

Andreae, B. (1972) [Types of farming in the tropics. Land use and livestock farming in the transition between tradition and progress.] Landwirtschaftliche Betriebsformen in den Tropen. Bodennutzung und Viehhaltung im Spannungsfeld von Tradition und Fortschritt. Hamburg, German Federal Republic; Paul Parey 190pp.

Andreae, B. (1974) [The farm sector at the agronomic boundaries of aridity.] Die Farmwirtschaft an den agronomischen Trockengrenzen. *Geographische Zeitschrift, Beihefte Erdkundliches Wissen* No.38 67pp.

Andreae, B. (1976) [Stages of development of ranch farms in arid zones of the continents.] Entwicklungsstufen der Ranchbetriebe in den Trockenzonen der Kontinente. *Tropenlandwirt* 77, 7–24.

Andreae, B. (1977) [Agricultural geography.] Agrargeographie. Berlin; New York, USA; Walter de Gruyter 332pp. [English language edition forthcoming, New York, USA; Walter de Gruyter 1980.]

Andreae, B. (1978) Means of increasing productivity in extensive grassland farming in arid areas of Africa. *Geo Journal* 2 (4), 331–342.

Bähr, J. (1970) [Structural change in the farm sector of Southwest Africa.] Strukturwandel der Farmwirtschaft in Südwestafrika. *Zeitschrift für Ausländische Landwirtschaft* 9, 147–159.

Bauer, H. (1972) [Thoughts on the fresh milk supply to towns in developing countries.] Gedanken zur städtischen Frischmilchversorgung in Entwicklungsländern. *Zeitschrift für Ausländische Landwirtschaft* 11, 318–334.

Binzer, R.; Kirchgessner, M.; Kiermeier, F.; Probst, A.; Bartha, R. (1971) [Feed supply in the Sahelian zone of Africa and its influence on milk yield characteristics.] Zur Futterversorgung in der sahelinen Zone Afrikas und ihre Einfluss auf Milchleistungseigenschaften. *Wirtschaftseigene Futter* 17 (2), 154–156.

Cole, M.M. (1961) South Africa. London, UK; Methuen & Co. Ltd. 696pp.

Gregory, S. (1969) Rainfall reliability. In: Thomas, M.F.; Whittington, G.W. (*Eds*) Environment and land use in Africa London, UK; Methuen & Co. Ltd. 57–82.

Hase, H.J. von (1964) [The effect of drought years in South West Africa and how they are overcome.] Die Auswirkungen de Dürrejahre in Südwestafrika und ihre Überwindung. *Deutsche Tropenlandwirt* 65 p.43.

Herzog, R. (1967) [Nomads' problems of adaptation.] Anpassungsprobleme der Nomaden. *Zeitschrift für Ausländische Landwirtschaft* 6, 1–21.

Huhn, J. (1977) [Nomadic herding at the arid limit of livestock farming.] Das Hirtennomadentum an den Trockengrenzen der Viehhaltung. In: Agrarentwicklung auf Grenzlandorten der Tropen. *Fachbereich Internationale Agrarentwicklung, Technische Universität Berlin* Studien 4 (17), 59–77.

Jentzsch, E.G. (1965) [Structure of food supply and agricultural production in Tunisia in the past, present and future.] Die Struktur der Nahrungsversorgung und der landwirtschaftlichen Produktion Tunesiens in Vergangenheit, Gegenwart und Zunkunft. Dissertation, Technische Universität Berlin 258pp.

Klimm, E. (1974) [Ngamiland. Geographical conditions and prospects for its economy.] Ngamiland. Geographische Voraussetzungen und Perspektiven seiner Wirtschaft. Kölner Geographische Arbeiten No.6 277pp.

Kmoch, H.G. [Development of feed production in the savannah areas of Africa.] Die Entwicklung der Futter produktion in den Savannengebieten Afrikas. *Arbeitsgemeinschaft für Forschung des Landes Nordrhein-Westfalen* 155 41–76.

Knapp, R. (1965) [Combination of plant types, development and natural productivity of pasture vegetation in arid areas in various climatic regions of the world.] Pflanzenarten-Zusammensetzung, Entwicklung und natürliche Produktivität der Weide-Vegetation in Trockengebieten in verschiedenen Klima-Bereichen der Erde. In: Weide-Wirtschaft in Trockengebieten. *Giessener Beiträge zur Entwicklungsforschung, Reihe I* 1, 71–97.

Manshard, W. (1965) [Livestock farming in arid areas of tropical Africa—a geographical survey.] Die Viehhaltung in den Trockengebieten Tropisch-Afrikas—ein geographischer Überblick. In: Weide-Wirtschaft in Trockengebieten. *Giessener Beiträge zur Entwicklungsforschung, Reihe I* 1, 29–36.

Meyn, K. (1970) Beef production in East Africa. *IFO-Forschungsberichte der Afrika-Studienstelle* No.29 227pp.

Schickele, R. (1931) [Investigations into forms of pasturing in arid areas of the world.] Untersuchungen über die Formen der Weidewirtschaft in den Trockengebieten der Erde. Dissertation, Landwirtschaftliche Hochschule, Universität Berlin 151pp.

Schinkel, H.G. (1970) [Livestock husbandry, breeding and care among nomads in East and Northeast Africa.] Haltung, Zucht und Pflege des Viehs bei den Nomaden Ost- und Nordostafrikas. Berlin, German Democratic Republic; Akademie-Verlag 302pp.

Schiffers, H. et al. (1976) [After the drought. The future of the Sahel.] Nach der Dürre. Die Zukunft des Sahel. *Afrika-Studien, IFO-Institut für Wirtschaftsforschung* No.94 384pp.

Stamp, L.D. (1965) A history of land use in arid regions. Paris, France UNESCO 388pp. [Arid Zone Research Vol.17, Ed.2].

Uhlig, H. (1965) [Geographical basis of pasturing in arid areas of the tropics and sub-tropics.] Die geographischen Grundlagen der Weidewirtschaft in den Trockengebieten der Tropen und Subtropen. In; Weidewirtschaft in Trockengebieten. *Giessener Beiträge zur Entwicklungsforschung, Reihe I* No.1 28pp.

USA, Department of Agriculture (1956) Costs and returns. Commercial family-operated farms by type and size 1930-1951. *Statistical Bulletin, Economic Research Service* No.197 67pp.

USA, Department of Agriculture (1969) Farm costs and returns. Commercial farms by type, size and location 1960–1969. *Agricultural Information Bulletin, Economic Research Service* 230, 79–97.

USA, Department of Agriculture (1971) Handbook of agricultural charts. *Agricultural Handbook* No.423 159pp.

USA, Department of Agriculture (1972) Indices of agricultural production in Africa and the Near East 1962–71. Economic Research Service-Foreign No.265 52pp.

Walter, H. (1939) [Grassland, savannah and the bush in the arid part of Africa and their ecological conditions.] Grasland, Savanne und Busch im ariden Teile Afrikas und ihre ökologische Bedingtheit. *Jahrbücher für Wissenschaftliche Botanik* 87.

Walter, H.; Volk, O.H. (1954) [Principles of pasturing in Southwest Africa.] Grundlagen der Weidewirtschaft in Südwestafrika. Stuttgart, German Federal Republic; Eugen Ulmer 281pp.

Webster, C.C.; Wilson, P.N. (1969) Agriculture in the tropics. London, UK; Longmans 488pp. Ed.3.

Whyte, R.O. (1967) Milk production in developing countries. London, UK; Faber; New York USA; Praeger 240pp.

Weigley, G. (1971) Tropical agriculture. The development of production. London, UK; Faber 376pp.

# Chapter 5

# Increasing the Productivity of Arable Rain-fed Farming in the Tropics

## Farming Rain-fed Regions (Dryland Arable Farming)

Of the six tropical climatic zones named in Chapter 3 only the last three and especially the humid and dry savannahs and various tropical uplands can be considerd for rain-fed farming. In all cases rain farming was originally based on the ancient system of shifting cultivation, in which a few years of arable farming are interspersed with a few or very many years of the original vegetation. Arable farming has developed and is still developing in completely different ways under different climatic conditions.

## Rain farming near the agronomic limit of aridity (dry Savannah)

Rain farming in the dry savannah (also called short-grass savannah) is influenced by atmospheric humidity, solar radiation, wind speed, type of crops, etc., but in general begins wherever the rainfall level is 250 to 400 mm (cf.Fig.8) in Africa, 300 mm in Iran and 350 to 400 mm in Arabia and Central Asia. However, where the rainfall distribution is particularly disadvantageous, this agronomic limit of aridity can be associated with as much as 600 mm of rain annually, as in parts of East Africa.

Farming here originated with the shifting-cultivation system which also formed the basis of two thousand years of development in large parts of Europe. Referring to this cultivation system, C. Tacitus wrote in his "Germania": "Arva per annos mutant et super est ager". According to Aereboe, this is to be understood as follows: "The cornfields change each year and there is enough land left for arable use".

As long as the density of settlement is sparse and the land considered for arable use is predominantly natural grassland, the farmers break up a small piece of the grassland with a digging stick or hoe (or later with a plough). They then plant crops (mainly millet) on it for two to four years, until yields become very low owing to the lack of thorough soil cultivation and an absence of any kind of manuring. The land is then left again to revert to its natural grass cover for as many years as are needed for its soil fertility to be restored. Meanwhile another piece of land is taken instead and worked in the same way. Later on, the farmers may grow grain again on the pieces of land previously used but only at irregular intervals of many years.

The further development of the steppe or savannah shifting-cultivation system towards more productive forms of agriculture is made necessary by population increase. It is simplified by technical progress in tillage methods, fertiliser use, plant breeding etc., and is aimed primarily at increasing the proportion of land used productively.

## Table 9  Farm management characteristics of tropical crops.

| Crop | Height m OD | Rainfall mm/year | Mean temperature °C | Growing-period months | Peasant farms (B) Large farms (G) | Industrial processing | Irrigation | Labour input (d = Man-days/ha, h = Man-hours/ha) | Physical yields dt/ha[1] | Gross returns DM/ha | Special remarks |
|---|---|---|---|---|---|---|---|---|---|---|---|
| **A. CLIMATE IN THE TROPICAL RAIN FOREST** | | | | | | | | | | | |
| Cassava (Manioc) | up to 2000 | from 1500 | 20 | 7–24 | B | (X)* | – | 200–225d | 125–250 (63) | 470 | Raw material for tapioca flour |
| Yams | | from 1100 | 20–25 | 6–12 | B | – | – | 375–600d | 50–175 (71) | 560 | Typical subsistence crop |
| Sweet potatoes | up to 1500 | 500–1250 | 20–22 | 3–6 (8) | B | – | (X) | 175–200d | 75–150 (71) | 560 | Typical subsistence crop |
| Jute | lowlands | 1500–2500 | 24–35 | . | B | XX | – | 1 200h | 12.5–16.8 (10.5) (fibre) | 1 060 | Fibre content = 4.5–7.5% of harvested weight |
| **B. HUMID SAVANNAH** | | | | | | | | | | | |
| Rice (wet rice) | up to 1200 | over 800 | above 20 | 3–6 | B | (X) | X | 600–1 200h | 7–30 (19.2) | 200–400 | Mountain rice has lower yields |
| Maize | up to 2800 | 760–1500 | | 4–6 | B. G | XX | (X) | 400–860h | 20–50 (10.8) | 520 (Kenya) | Often the main food |
| Sugar cane | up to 1600 | from 1200 | 25–28 | 15 and more | B. G | XX | XX | 400–630h | 32–41 | 2000–3000 (Kenya) | Labour = 60% of costs |
| Bush beans | up to 1600 | | | 3–5 | B. G | – | (X) | | (6.0) | . | |
| Soya | Monsoon constitution | | very high | 80–200 days | B | XX | . | 80–85d | 12–22 (7.7) | 320–360 (Tanzania) | 18% fat. 14% protein |
| Castor oil plant | | 800–1000 | | perennial | . | XX | . | high | 10–40 | | Seed contains 40–60% oil of a very high viscosity |
| **C. TROPICAL HIGHLANDS** | | | | | | | | | | | |
| Tobacco | | 500–1000 | 21–27 | 3–6 | B. G | XX | (X) | 2000–7000h | ca. 10 (8.3) | 4000–12 000 | – |
| Potatoes | 1800–2500 | 1000–1800 | 15–18 | 4–7 | B | – | – | 1200h | 80–200 (70) | ca. 2000 (Kenya) | Long storage only possible at <10°C and then difficult |
| Pyrethrum | 2100–2600 | 1000–1500 | | mostly 3 year crops | B. G | XX | – | 600h | 5–9/year | ca. 1400 (Kenya) | Flowers are raw material for insecticide |
| Teff (Eragrostis abyssinica) | 1300–3000 | slight | | short | B | – | – | . | 4–9 | | For flat loaves of bread. Straw rich in nutrients (Dry period), especially Ethiopia |
| Nug (Guizotia olifera) | | | | | | | | | | | – |
| | Grown particularly in Ethiopia. Seed contains 36–43% fat. Oil does not go rancid. High resistance to weeds. | | | | | | | | | | |
| **D. ARID TROPICS** | | | | | | | | | | | |
| Millet (sorghum) | up to 2500 | 380–1300 | 32–35 | 2–6 | B | – | – | 30d | 3–16 (6.6) | 180–220 | Often main food |
| Chick peas | . | drought resistant | wine climate | . | B | – | – | . | 5–7 (6.3) | . | Third most important pulse crop in the world (after soya and phaseolus) |
| Ground nuts | . | 300–1200 | 22 | 3–5 | B | X | (X) | 43–130d | 6–17 (8.1) | 300–600 | Dry period necessary for harvest |
| Sesame | all over the tropics | | 25 | 4 | B | X | – | 55–60d | 3–10 (2.9) | 220–260 (Tanzania) | Seed 50–60% oil content |
| Cotton (short staple) | . | 500–1500 | 26–28 | 4–7 | B. G | XX | X (X) | 1 450–1 550h | 4.0–6.5 (4.6) | ca. 480 (India) | High harvest labour peak |
| Kenaf | . | 500–650 | . | 4–5 | B | . | . | 54.4d | 17–22 (fibre) | . | Fibre in stem 16–18% |
| Sunflower | Dry places with warm summers, dry weather in the ripening periods | | . | . | B (G) | X | . | | 6.0 (oil) | 250 | |

[1] From Klammern's "Average yields in Africa", based on the FAO Production Year Book Vol. 23 (1969) Rome 1970, compiled mainly from Frank et al. (1967).

* X = desirable; XX = necessary.

Fig. 14
The northern limit of rainfed farming in Chad. (Source: Statistische Bundesamt 1972, p.5)

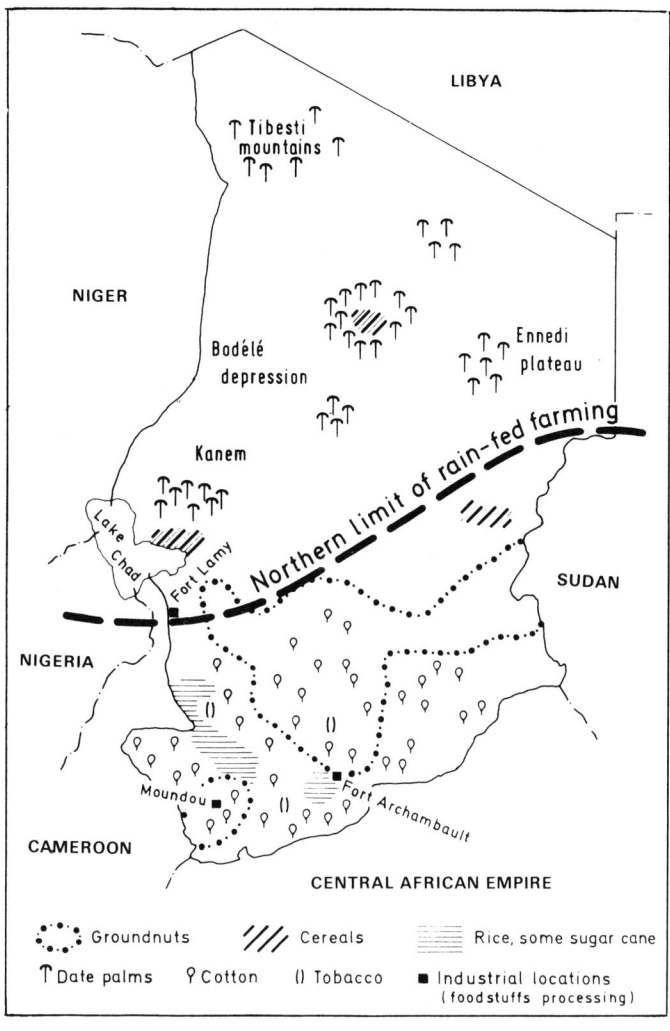

If three years of cultivation follow 15 years of grass-fallow, the proportion of land used is only one sixth. However, if four years of arable alternate with 12 years of grassland, the proportion increases to one quarter. Finally, if eight years of grass are followed by four years of arable a figure of one third is reached. A precondition for this sort of development is, however, that the natural grassland's effect of increasing fertility is replaced by other measures to encourage fertility. This is not easy in the dry savannah.

The more the grass-fallow is reduced, the more water shortage becomes a limiting factor and determines the farm organisation. The farmer faces two alternatives. He must be content with crops which need little water but whose yields are low, but where he can be sure of obtaining an annual harvest. Alternatively, as in the dry farming system, he may choose crops which need more water but which also yield more, and at the same time accept the need for fallow areas to store water.

If the first method is used crops needing little water are those such as millet, ground-nuts, Bechuana beans, chick peas, sesame and sisal. This explains the very basic importance of millet growing for all the arid regions of Africa and also the predominance of ground-nuts in the states bordering the Sahara, such as Niger, Chad, Upper Volta, Mali or Senegal. Sisal comprised 70.5% of Tanzania's total agricultural exports in 1976, whilst ground-nuts and ground-nut products accounted for 56.5% of Niger's agricultural exports in 1976.

Fig. 15
Zones of intensity of the dry farming system

| Rainfall mm / year | Long Rainy season | Short | Fallow % arable land |
|---|---|---|---|
| ca. 300 mm | (A)<br>1. Bare fallow<br>2. Wheat | (B)<br>1. Bare fallow<br>2. Barley (four-row) | 50 % |
| ca. 350 mm | (C)<br>1. Bare fallow<br>2. Wheat<br>3. Millet | (D)<br>1. Bare fallow<br>2. Groundnuts<br>3. Millet | 33 % |
| ca. 400 mm | (E)<br>1. Bare fallow<br>2. Wheat<br>3. Wheat | | 33 % |
| 450 – 500 mm | (F)<br>1. Bare fallow<br>2. Wheat<br>3. Maize, Blue lupins<br>4. Wheat | (G)<br>1. Bare fallow<br>2. Groundnuts<br>3. Sorghum, Millet<br>4. Millet | 25 % |
| Transition to humid savannah | (H)<br>1. Sunflowers, Groundnuts<br>2. Maize | (J)<br>1. Dried peas<br>2. Wheat | —— |

The second method, the dry farming system, generally demands a higher level of technology, and is therefore easier to practise in more industrialised states. Thus the main areas using the system are the arid regions of the U.S.A. and Canada, and of Argentina and Australia, the Republic of South Africa and the USSR. This system is also found in the northern Sahara border states and in Iran and Turkey, as well as in many other countries. As Fig. 15 shows the main crops are wheat and barley though cotton is also grown sometimes.

In a very dry climate (about 300 mm of rain) with a crop rotation of bare fallow followed by wheat (or with a shorter rainy season, bare fallow followed by summer barley) half the arable area has to be devoted to fallow. With increasing rainfall, the proportion of fallow can be cut back, although soil conserving crops (apart from ground-nuts which need very little water) are introduced only slowly. The fallow serves to increase the yield of the following cereal crop, primarily by saving water but also by releasing nutrients. The whole principle on which the dry farming system is organised is based on the fact that bare soil evaporates less water than plant cover so that some of the rain falling in the fallow year can benefit the following cereal crop.

## Rain farming in the humid savannah

The humid or high-grass savannah (also called the savannah forest zone) is the climatic area of the tropics where rain farming is most important. In more arid zones it is defined by grassland, and in more humid ones by tree and shrub cultivation.

The predominant short-life field crops in the humid savannah are maize, rice, phaseolus, groundnuts, and others. Ecological conditions here are also acceptable for cassava and yams though not yet optimal. Table 10 shows some examples of crop rotations:

Table 10   Crop rotations for rain farming in the humid tropics.

| A. West Africa | B. Malawi | C. India |
|---|---|---|
| 1–15. forest fallow | 1. groundnuts | 1. sugar cane |
| 16.  hill rice | 2. cotton | 2. vegetables |
| 17.  beans, yucca | 3. rice | 3. rice |
| 18.  cassava | | |

In Africa and India, mixed cropping is very important on small farms. Examples given by Könnecke (1967) for West Africa are yams and guinea corn, millet and guinea corn, maize and oil pumpkin, hill rice and cotton, and maize and groundnuts. For India examples are cotton and Italian millet, bush peas and Italian millet, and cotton and coriander (spice plant). Ethiopia has additional important crops in its upland regions such as teff (Eragrostis abyssinica), from which the popular flat loaves (indjera) are made and whose straw is rich in nutrients and makes a welcome fodder in the dry season. Another useful crop is the oil-producing plant nug (Guizotia olifera) which is excellent for combating weeds and whose oil does not become rancid even under primitive storage conditions.

Where the climate of the humid savannah has some cool months, as in the tropical uplands, crops which prefer moderate climates are grown (e.g. wheat, barley, vegetables, potatoes). Wheat is grown in Kenya at 2000 m above sea level and in Ethiopia, which is further from the equator, at 1700 m above sea level. All African wheatlands also produce barley, but the reverse is not true. In the countries like those between the Sahara and the Mediterranean where the rainy season is shorter and the climate drier wheat is outstripped by barley because of barley's special properties of a short life cycle and resistance to drought. However, these areas can no longer be considered humid savannah.

In the humid savannah the ancient system of shifting cultivation is again the basis of field systems. According to the natural vegetation there is an alternation of forest and arable, or grass and arable. The incentives and aims to progress from this system are the same as for the dry savannah but the measures to be taken

Fig. 16

Ley farming in African humid savannahs. 1) Source: Niederstücke 1970, p.76; 2) 6 ha of agricultural land and 4 workers per farm; 3) 18 ha agricultural land, 600 mm rain, dairy cattle, 0.55 ha principal feed-crops/cattle unit, gross return DM2 540 per ha agricultural land, net return DM1 000 per ha)

### Kenya Highlands (Kericho District)[1]

| Large farms | | Small farms | |
|---|---|---|---|
| A | B | C | D |
| 1.-3. Cultivated grass | 1.-3. Cultivated grass | 1.-4. Natural grass | 1.-5. Natural grass |
| 4. Potatoes - Beans | 4. Potatoes - Beans | 5. English potatoes- Vegetables | 6.-8. Maize |
| 5.-7. Pyrethrum | 5.-6. Wheat | 6.-9. Maize | 9. Sorghum, millet |
| 8. Maize | 7. Barley | | 10. Sweet potatoes- Vegetables |
| Grassland % of arable land | | | |
| 38 | 43 | 44 | 50 |

### South-East Africa

| Rain-fed farming | | Irrigated farming |
|---|---|---|
| E | F | G |
| Malawi [2] | North-Transvaal | Mid-Transvaal (Rustenburg) [3] |
| 1.-3. Cultivated grass | 1.-3. Cultivated grass | 1.-3. Lucerne |
| 4. Tobacco | 4. Sorghum, millet | 4. Maize silage |
| 5. Cotton | 5. Sorghum, millet | 5. Summer: Grass silage;  Winter: Green oats |
| 6. Groundnuts | 6. Bechuana beans | 6. Summer: Bean hay;  Winter: Wheat |
| 7. Cotton | | 7. Summer: Tobacco;  Winter: Green oats |
| 8.-9. Maize | | 8. Summer: Tobacco;  Winter: Wheat |
| | | 9. Summer: Maize silage;  Winter: Wheat |
| Cultivated grass or lucerne % of arable land | | |
| 33 | 50 | 33 |

differ. The central problem here is not the economic management of scarce ground-water, but finding a method of maintaining soil fertility which does not involve long years of natural vegetation. It is necessary to combat losses of humus without using traditional methods of years of forest or grass fallow, to limit erosion damage, at least to moderate the loss of nutrient content, to master weed growth and to prevent disasters caused by plant diseases and pests. All this can be most successfully achieved by maintaining plant cover throughout the year, but it is precisely this which is prevented by the dry season.

As the population grows, an increasingly large proportion of the usable land must be cultivated and the areas available for grass or forest fallow diminish. Thus in the humid savannah, the best replacement for fallow under the original natural vegetation is a cultivated grass crop. Examples are shown in Fig. 16.

The percentage of land on which grass has to be grown to conserve the soil is considerable and it is mostly far in excess of the amount that can be used for fodder. Ley farming is technically possible but at the beginning it is an economically-expensive rotation system. As long as the market for milk and meat in developing countries remains poor and the communications system is incomplete part of the fodder growth cannot be utilized. The total cost of the grass crop must therefore be considered as the cost of promoting soil fertility. This situation will eventually change and there is no doubt that time will operate in favour of the ley farming system of soil conservation in the humid savannah.

### Rain farming near the agronomic humidity limit (tropical rain forest climate)

There are not very many short-lived crops which prefer the hot and permanently wet climate of the tropical rain forests. Root crops such as manioc (cassava), sweet potatoes and yams are grown, but these are typical crops of subsistence farmers. As far as marketable grain crops are concerned, rice, with appropriate water supplies, and maize, because of its wide ecological distribution, are possibilities. Sugar cane grows well here. Once we progress beyond the still widespread system of shifting cultivation by subsistence farmers the margins to be made from rain-fed farming for the market are not very great. Above all, once one ceases to use the system of forest fallow for regenerating the soil, the soil fertility situation in the tropical rain forest area is even more precarious than in the humid savannah.

## Basic Difficulties Facing Agriculture in the Humid Tropics

### Natural obstacles to agriculture

The humid tropics are defined as the climatic zone near the equator whose natural vegetation consists of dense virgin forest. This covers all those regions where the dry season is less than four months and where the annual precipitation is at least 1200 mm. Wherever nature alone determines the vegetation in these regions without interference by man, we find permanent mixed plant associations. The biological equilibrium which guarantees the preservation of soil fertility in the humid tropics and which nature is able to preserve far more effectively than any human planning could do, depends on: (1) a well-balanced plant association, (2) the prevalence of perennial plants within this association, and (3) the presence of trees and shrubs within this plant association. Nature wants it this way.

The problem of soil fertility thus differs in scope and direction, depending on the crops and cultivation methods used by man to utilize the land and on the extent to which these methods clash with these three basic natural principles. The mixed cultivation of trees and shrubs is best adjusted to the natural vegetation and in general does not affect the fertility of the soil. When trees and shrubs are planted in a one-crop system, for instance on cocoa, rubber or banana plantations, the soil fertility problem does arise. It assumes more serious proportions under any form of grassland utilization. The fertility of the soil is most seriously harmed under short-lived field crops.

Field crops are often cultivated in monoculture; i.e. in a one-crop system using mostly short-lived crops. After such crops are harvested, the soil is deprived of its protective vegetation cover and is exposed to rain and solar radiation. In the humid tropics such field cropping is a form of land utilization which is opposed to nature's own way. One-year one-crop systems are detrimental to soil fertility unless man takes proper precautions to eliminate the damage they do to nature. For this reason the question as to how to preserve the fertility of the soil in humid tropical regions has occupied mankind as long as agriculutre has been carried on. In recent times this question has become particularly urgent as the population explosion has forced mankind to adopt more productive methods of land utilization and indeed to constantly extend the land under crops.

**Output-input ratios over several years of cropping**

The preservation of soil fertility is much easier in the temperate zone than in the humid tropics and much easier in industrial countries than in the densely populated agricultural countries. In Europe markedly higher crop yields are obtained than in the African tropics. The incomparably higher fertilizer input in Europe is only one of the many causes (see Table 11).

Table 11   Rice yields, nitrogen consumption and irrigation in some developing and developed countries.

| Country | Rice yields (paddy) dt/ha 1976 | Irrigation in % of arable land and permanent crops 1975 | Nitrogen in kg/ha N to arable land and permanent crops 1974 |
|---|---|---|---|
| Brazil | 14.5 | 2.6 | 10.6 |
| Madagascar | 17.3 | 14.9 | 1.1 |
| Philippines | 18.1 | 17.6 | 22.5 |
| Thailand | 18.2 | 19.0 | 4.8 |
| Burma | 18.2 | 9.4 | 3.7 |
| India | 18.3 | 19.3 | 10.6 |
| Pakistan | 23.2 | 73.5 | 18.5 |
| Indonesia | 26.1 | 23.5 | 21.6 |
| France | 40.9 | 3.0 | 48.0[1] |
| USA | 52.4 | 7.9 | 37.3 |
| Italy | 54.2 | 29.2 | 39.0[1] |
| Japan | 55.0 | 48.0 | 123.9 |
| Developing Countries | 19.7 | 13.5 | 10.7 |
| Developed Countries | 54.3 | 7.6 | 40.9 |
| World | 24.5 | 15.1 | 28.7[2] |

[1] 1974/75; [2] 1975/76.

*Note:* In Southeast Asia every kg N increases the rice yield about 12–30 kg. In the Philippines the price of 1 kg fertilizer is equivalent to 3.3 kg rice. Nitrogen input is therefore highly profitable.

Source: F.A.O. (1976 1977).

In addition, the proportion of arable land cultivated every year is considerably higher in the temperate zone than it is in the humid tropics. In Europe such annual cropping has been carried on for roughly a century and a half and yet crop yields have constantly increased. On the other hand, in the humid tropics short-term cropping is followed by long-term fallowing under natural vegetation, so that the proportion of crop land available every year can only be very small. Even so, yields decline sharply in the course of a short continuous cultivation period of two to three or at the most four years. Fig. 17 illustrates this decline in yields under prolonged cropping in a system of shifting cultivation in the humid tropics.

Fig. 17
Decline in yields under prolonged cropping in the humid tropics with a system of shifting cultivation. (Source: Nye & Greenland 1960)

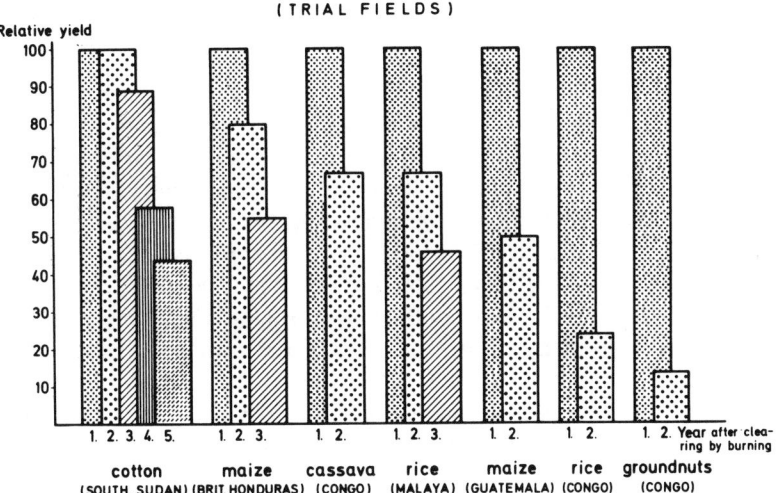

The constant and often rapid deterioration of soil fertility under cropping is an almost universal characteristic of tropical soils. In addition, weed growth and tillage resistance rapidly increase in cropping over several years so that under prolonged cropping yields per workday decrease even more rapidly than those per ha of cropland. The output-input ratio becomes progressively more unfavourable, and after three or four years reaches the point where further continuation of cropping becomes economically unprofitable.

## The causes of the decline in yields

In this context only a very brief description can be given of the reasons for the rapid decline in yields. This is, however, sufficient for examining the economic position (see also Vageler 1942).

In temperate zones the effective components of the soil include clay and humified organic matter. These substances not only produce the desirable crumb structure but also condition the absorptive capacity of the soil. They store water and nutrient elements in the soil and release them to the plants when needed. The humus content of virgin soils untouched by man is in equilibrium with the climatic conditions, i.e. nature has balanced the decomposition and re-formation of humus. In arctic regions where all living matter is buried under ice and snow throughout the major part of the year, only a very small amount of vegetable matter is formed but an equally small amount is decomposed. At the other extreme the decomposition of organic matter in equatorial regions is between five and ten times higher than in temperate zones, but the rich natural flora balance this loss as the tropical rain forest produces between 12 and 18 tons of dry matter per ha. If man destroys this rich natural flora, the balance of organic matter is disturbed.

In the past temperate zone farmers concentrated part of their efforts on preserving or producing a stable organic soil content at as high a level as possible. The high and stable yields of field crops in Central Europe are largely due to high dressings with animal manure and green manure as well as to suitable tillage and other measures. If wrong cultivation methods lead to a loss of organic matter, crop yields will decrease and costs will increase even in a temperate climate, but this will not lead to a collapse of the entire farming system. Even if the organic resources of the soil are largely exhausted, the soil fertility in Central Europe cannot fall below certain minimum levels. This is firstly because with a lower degree of weathering the power of natural mineral resources to supply nutrients is reasonably high. Secondly it is because naturally released nutrients or those added as fertilizers are almost completely absorbed by clay minerals which usually have a fairly high absorption capacity for nutrients and water (sandy soils are the exception).

The case in the humid tropics is quite different. The clay minerals prevailing in many of the highly weathered, red latosolic soils have only a small cation exchange capacity. Thus, although they have a high clay content, their adsorption capacity and even more the power of these tropical soils to supply nutrients depends almost entirely on their organic content. The key problem of fertility in tropical soils is, therefore, to produce and to preserve a certain content of organic matter.

This is, in fact, rather difficult because the decomposition of humus following the clearing of land in the humid tropics is much more rapid and intense than in temperate zones. The absence of vegetative cover after clearing leads to a rapid mineralization of the organic matter due to intensive solar radiation and aeration of the soil. Land cleared of virgin forest loses 95% of its humus down to a depth of 40 to 50 cm in about one year. The decomposition of organic matter can hardly be stopped. The tropical soil deprived of its humus content, although of high clay content, lacks cohesion and plasticity and it acquires the characteristics of a sandy soil due to a stable aggregation by a high content of sesquioxides. Such soil is completely "run down". Someone once used the metaphor that it resembles a sieve which allows rainwater and nutrients to run through to the underground rapidly and unimpeded without enriching the soil. A rapid decline in yields is the inevitable result.

## Shifting Cultivation: the Ideal Solution for Agriculture in the Humid Tropics

Since the highly weathered soil in the humid tropics lacks sufficient storage and supply capacity for nutrients, the nutrient elements must be stored in plant material.

This explains why over thousands of years people in the tropics have adhered to shifting cultivation and why, even today, the system is still carried on by over 200 million people on over 30 million square kilometres. As long as the economic conditions make shifting cultivation possible, it is indeed the ideal solution for agriculture in these conditions.

## Characteristics of the system

This system consists of the clearing of virgin forest by fire and the cultivation of crops on cleared areas for two or three years, or at the most four. As soon as yields cease to be satisfactory the cleared areas are abandoned to revert to natural forest as fallow for many years or even decades and to regenerate the fertility of the soil. Virgin forest which increases fertility alternates with cropping which consumes fertility. Man leaves it to nature to repair the damage caused by growing annual crops which require a much higher supply of nutrients than the natural perennial plants. Table 12 gives some African examples of shifting cultivation.

Table 12   Examples of shifting cultivation in the humid tropics of Africa.

| Country or region | Rain in mm/year | Field crops | Virgin vegetation | No. of cropping years | No. of fallow years | Land under crops in % of available land |
|---|---|---|---|---|---|---|
| Zambia | c. 1125 | . | Miombo dry forest | 2 | up to 25 | 7 and more |
| Liberia | 2000–4500 | Rice, cassava | Rain forest | 1–2 | 8–15 | 11–12 |
| Sierra Leone | 2250–3250 | Rice, cassava | Rain forest | 1.5 | 8 | 16 |
| Central Congo | 1750 and more | Rice, maize, cassava | Rain forest | 2–3 | 10–15 | 17 |
| Nigeria, Umuahia | c. 2250 | Maize, yams, cassava | Bushland (Acioa barteri) | 1.5 | 4–7 | 18–27 |
| Abeokuta | c. 1250 | . | Thicket | 2 | 6–7 | 22–25 |
| West Africa | 1500–2000 | Maize, cassava | Moist semi-deciduous forest | 2–4 | 6–12 | 25 |
| Zambia | c. 1250 | | Thicket | 6–12 | 6–12 | 50 |

Source: Nye & Greenland (1960).

The number of years cropping depends not only on the soil and climate but also to a considerable extent on the predominant crops. Thus it increases as the climate becomes drier and as grain crops tend to predominate over root and tuber crops. The number of fallow years required for the restoration of a high level of soil fertility will largely depend on the amount of rainfall and the natural vegetation that this produces. Although the sequence of forest fallow—bush fallow—grass fallow accelerates the regeneration of the soil, it renders it increasingly imperfect. The original and natural level of soil fertility is completely restored only by long-term forest fallow. The number of fallow years given in Table 12, however, reflects the minimum period of rest under fallow which is practicable in the light of population density rather than the time required for a complete restoration of the natural fertility of the soil.

### The soil-regenerating functions of the system

The process of soil regeneration under forest fallow is mainly effected by the regrowth of organic matter, partly in the form of the densely growing vegetation and partly through a better conservation of soil organic matter by the shading of the soil surface and the resultant reduction of soil temperature and aeration. Large amounts of plant nutrients are stored in the natural forest fallow. The topsoil is also enriched with plant nutrients. As the roots reach down to the deeper subsoil, often to a depth of seven and more metres, natural nutrients from the subsoil as well as part of the nutrients which had been leached out are transported from the deep subsoil to the soil surface via the leaf litter. In addition the dense natural vegetation destroys weeds, plant diseases and pests which have spread during the cropping period.

The dense virgin forest is itself a gigantic reservoir storing large amounts of plant nutrients in its vegetation rather than in the soil. The nitrogen contained in the trees amounts to roughly 40% of the nitrogen content of the root zone. The amount of potassium stored in the vegetation cover roughly corresponds to the potassium reserves absorbed in the soil. In the case of phosphorous, as much as 10 times the amount of the reserves absorbed in the soil are stored in the living organic matter. (Welte 1963 p.226).

Clearing by burning is nothing more than a violent and rapid mobilization of these reserves of soil fertility. This mobilization takes place in two successive phases. The first phase is the actual burning. This converts the nutrient elements stored in the above-surface vegetation matter into minerals in the form of ash and so makes them accessible for the production of field crops. However, most of the nitrogen is lost in the process of clearing by burning, and potassium is particularly susceptible to leaching in the following rainy season. The high crop yield obtained in the first year of cropping is largely due to this direct effect of plant ash.

The second phase is the decomposition of organic matter including roots in the top soil which sets in intensively after burning. These mineralized nutrients, including nitrogen, are now available for the production of field crops. The harvest of the second, the third and possibly the fourth year of cropping are largely the effect of this decomposition process. The decline in yields if the cropping period is further prolonged is due to the fact that this process slackens and comes almost to a standstill after a few years. Under the system of shifting cultivation a reasonable sequence of nutrient utilization during the cropping period alternating with nutrient storage during the fallow period permits effective agriculture.

### The economic requirements and limitations of the system

The system of shifting cultivation can be maintained with a minimum input of capital. Yield-increasing capital goods such as mineral fertilizers, plant protection products and weed control agents are replaced by the soil-regenerating power of the forest fallow. It is not possible to use draught animals or larger tools and machinery for tillage, soil treatment and harvesting because the cleared areas are strewn with tree stumps, roots and trees that have resisted the fire and these must not be removed if the forest is to regrow rapidly later. For this reason hoeing is the usual technique applied in shifting cultivation, and the capital investment required is confined to axe, cutlass and hoe. It is, in other words, practically zero.

The labour input is also very low, in particular because the clearing is largely done by the fire. Tillage may be kept to an absolute minimum because the forest fallow makes the soil loose and friable. It must, moreover, be kept to such a minimum if the oxidation processes by which organic matter decomposes are to be prevented rather than supported. Weed control is also negligible, because it is largely effected by the alternation of cropping and fallowing. The only basic requirement is a large area of land suitable for cropping because only a fraction of the available land is actually cultivated. As a rule the proportion is between 15–20%.

In sparsely populated agricultural countries land is available in abundance. On the other hand, the number of workers available is limited and all forms of capital goods are extremely scarce and expensive. However, the shortage of these production factors presents no problem when we consider the level of input of these factors required in shifting cultivation. Thus this system is not only ideally suited to the natural conditions of agricultural production in the humid tropics but it is also perfectly suited to the economic conditions of sparsely populated agricultural countries. However, if at a later stage, the population increases and land becomes increasingly scarce, the system will eventually cease to provide sufficient fertile soil. It can only be maintained as long as enough land is available in the form of virgin forest for shifting cultivation. In the evergreen rain forest the land available for shifting must cover at least four times the area of the land under cultivation so that two to three years of cropping may be followed by 10 to 12 years of rest under forest fallow. Thus a family of 10 persons who require 3 ha of land for cropping must have over 15 ha of land available to them, and a family requiring 4 ha of crop land must have over 20 ha of land at their disposal. Population growth leads to a disastrous and vicious circle, the fallow period is shortened to gain more of the crop land needed to ensure adequate basic nutrition. Abbreviated forest fallow, however, leads to poorer secondary forest with incomplete regeneration of soil fertility. This leads to lower crop yields. This in turn requires a further extension of the land under cultivation at the expense of the land under fallow, and leads to a further reduction in the fallow period, and so on.

In countries with a growing population the system of shifting cultivation thus proves to be a bottleneck from which there seems to be no way out. Even on fertile soils of recent volcanic origin in the Isle of Java this system can at the most provide food for 40 to 50 inhabitants per sq km (Finck 1963 p.116), (e.g. the present population density in parts of East Nigeria). On less fertile soils in Africa the critical margin is much lower, and often lies between 30 and 35 inhabitants per sq km, (e.g. the present population density in Tunisia or Fiji). Even countries like Kenya with 23 inhabitants per sq km or Tanzania with 16 inhabitants per sq km are already complaining about a shortage of land because most of their soils have a low level of natural fertility. If population increases beyond the food production capacity of soil under shifting cultivation food supplies are no longer adequate. The area under cultivation must constantly be expanded at the cost of the forest fallow and sooner or later the stage of permanent cropping must be reached. This, however, raises difficult problems in humid tropical regions. We have not yet found an absolutely reliable means of replacing the soil regenerating function of natural fallow in such regions. The following section therefore tentatively explores possible solutions to this problem of vital importance to the humid tropics.

## Possibilities and Limits to Improving Shifting Cultivation in the Humid Tropics

### Transition from uncontrolled to controlled shifting cultivation

The possibilities of improving shifting cultivation in terms of increasing soil productivity are soon exhausted and the limits are quickly reached. The traditional form of shifting cultivation is uncontrolled shifting where the period of cropping and the period of fallow are not fixed. This irregularity reduces the productivity of the soil.

Fig. 18
Crop yields under shifting cultivation in relation to the cropping interval. (Source: Ruthenberg 1965)

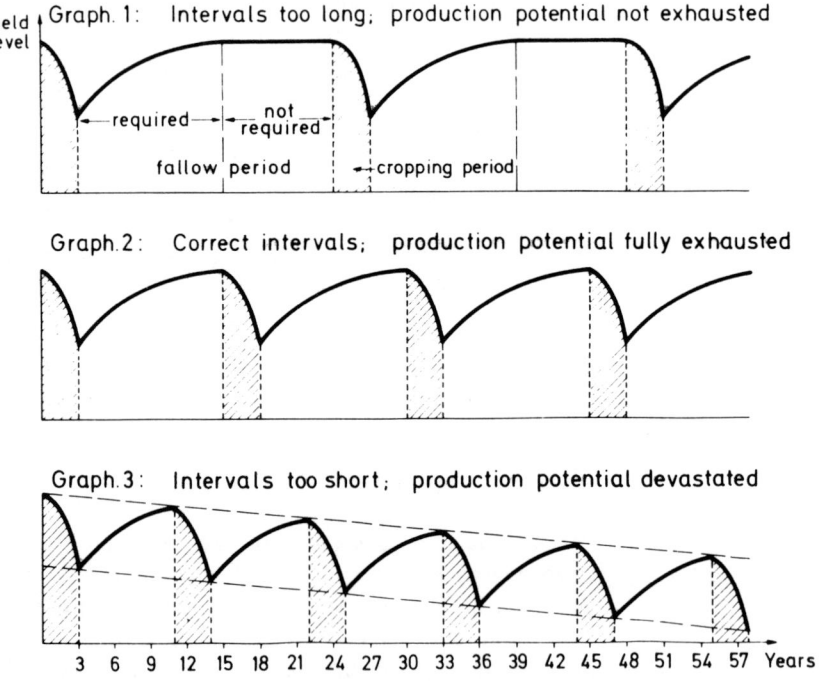

Fig. 18 shows how yields under shifting cultivation depend on the cropping interval. In graph 1 the fallow period is longer than necessary for the regeneration of soil fertility and production reserves are not fully utilized. In graph 3 the fallow period ends before soil fertility has returned to its original level so that production potential not only falls within each crop period but decreases cumulatively from cropping period to cropping period. This must also reduce the productivity of the soil significantly. Graph 2 shows the only case where cropping and fallow periods are adjusted so that soil fertility is completely restored and no land is wasted under unnecessarily long fallow. This is the only case which maximizes the productivity of the soil within the limits set by the system. The first step towards improving the system of shifting cultivation must therefore be to determine the exact number of years of cropping and of fallow required. In addition, a definite crop rotation plan must be drawn up for every region to ensure that the natural fallow is given the full time necessary for natural regeneration. Thus, this transition from uncontrolled to controlled shifting cultivation, presupposes a certain measure of regional planning and permanent settlements. Each settlement is allotted a certain area of land which is subdivided into the same number of felling strips as the total number years in the whole crop rotation plan, i.e., the total number of cropping years added to the fallow years. Fig. 19 shows such a scheme of controlled shifting cultivation based on the "corridor system". This system also known as "paysannats indigènes" has been successfully introduced in the Congo.

Fig. 19
Controlled shifting cultivation based on the corridor system. (Source: Dumont 1957 p.40)

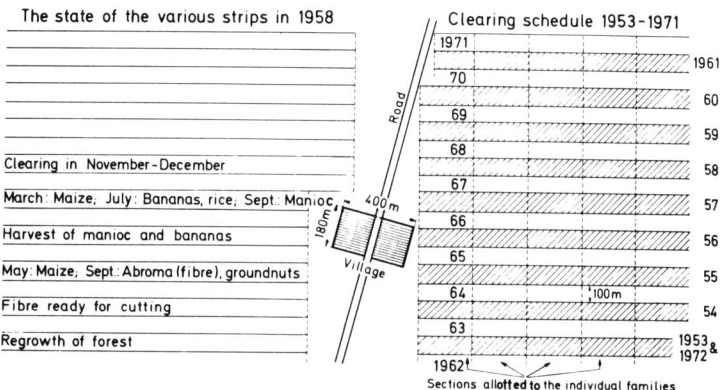

## Transition from natural to planted forest fallow

The second step in improving shifting cultivation is the transition from natural to planted forest fallows. Fast densely growing tree and shrub species appropriate to climate and soil should be chosen when planting such fallow. The young plants should be grown in nurseries till they are sufficiently strong and resistant and transplanted at the beginning of the fallow period. This technique should lead to a fast dense regrowth. The trees and shrubs store sufficient plant nutrients to accelerate the regeneration of soil fertility, shorten the fallow period and increase the productivity of the soil. Such a planted secondary forest naturally requires a considerable extra input of labour but this cost has to be borne for the sake of economising in land. This is a part of the economic framework of necessities and possibilities as a growing population increases the pressure on land and reduces the area available per caput.

## Transition from hoeing to ploughing

The system of artificially planted forest fallows makes the next big technological leap possible, the change from hoeing to ploughing. A natural regrowth of the forest is no longer wanted and the remaining living roots and stumps which are no longer required, may be cleared mechanically. This is an essential preliminary to ploughing and tillage on a wider scale with draught animals or mechanical equipment. The additional labour for planting the fallow is then offset by the reduced labour for clearing, cultivation and harvesting. The bullocks provide the natural manure necessary for higher crop yields. Considerably larger areas of land can be cultivated by ploughing than by hoeing and yields per ha are higher with the same number of workers. The transition from hoeing to ploughing raises the productivity of both soil and labour considerably.

## Possibilities of Replacing Shifting Cultivation

These are more or less all the possibilities for increasing productivity under shifting cultivation. If population continues to grow the system must be completely replaced by a new system of cultivation and so far this has rarely been done in practice.

### The transition from forest to leguminous fallow

If the ratio of cropping to fallowing has to be reduced below the margin which can be tolerated under controlled shifting cultivation even with planted forest fallow and ploughing the next logical step is the replacement of forest fallow by leguminous fallow. The soil regains its fertility in a shorter period under short-lived crops than under forest or brush. Grass crops absorb too much nitrogen to be economic at this stage of general economic development and their value as a preceding crop is too low. Leguminous crops are chosen because they absorb nitrogen from the air, hardly need mineral fertilizers, provide good shade and penetrate the soil deeply with their roots. In contrast to the case with forest burning the nitrogen stored by these crops is preserved and is available to the following crop. Depending on the local soil climate, fertility can be preserved by two years of leguminous fallow followed by one year of cropping, by one year of leguminous fallow with one of cropping or by one year of leguminous fallow with two years of cropping. The rate of land utilization is 33% in the first case, 50% in the second and as much as 67% of the available area in the third case. As table 13 shows this is considerably better than the rate of utilization under shifting cultivation. It is mainly achieved by a greater use of the manpower made available by the growth of the population. The input of capital continues to be low. It is confined to the necessary leguminous crop seeds in addition to the bullocks and equipment already procured at the earlier stage of cultivation.

### Transition from leguminous to grass fallow

More productive forms of cultivation can only be introduced when the economy has reached a stage of development which permits an increased input of capital. The prices of agricultural products must have risen and those of mineral fertilizers have fallen sufficiently to make it profitable to apply fertilizers. The more fertilizers are used the less important becomes the role of organic matter as a source of plant nutrients. The main function of organic matter then becomes erosion control through aggregation, increasing soil permeability and increasing available moisture. The application of larger amounts of fertilizer makes the transition from leguminous fallow to grass fallow possible. This produces a denser and more even root growth in the topsoil. The rapid growth of a dense root system stimulated by the application of mineral fertilizer accelerates the matting of the A horizon and promotes the build-up of humus in a less well aerated (oxidative) surface soil. The regeneration of soil fertility is accelerated and completed. The result is higher crop yields per ha.

Under favourable conditions the rate of land utilization once again rises. Depending on soil and climate in the area and the intensity of fertilizer application, the ratio of grass fallow to cropping years is 3:3 or 3:4 or even 2:4. Thus the rate of land utilization is 50%, 57% or even 67% of the available area of land (cf. Table 13).

Table 13 Approximate increase in land utilization in the evolution from shifting cultivation to more productive systems.

| Cropping system | Number of years | | Regrowth of forest or grass | | Land under cultivation as a percentage of available land |
|---|---|---|---|---|---|
| | cropping | fallow | natural | cultivated | |
| I. Long-term fallow vegetation | | | | | |
| Uncontrolled shifting cultivation | 2 | 26 | X | . | 8 |
| Uncontrolled shifting cultivation | 3 | 17 | X | . | 15 |
| Controlled shifting cultivation | 3 | 12 | X | . | 20 |
| Controlled shifting cultivation | 3 | 9 | . | X | 25 |
| II. Short-term fallow cultivation | | | | | |
| Leguminous fallowing | 1 | 2 | . | X | 33 |
| Leguminous fallowing | 1 | 1 | . | X | 50 |
| Grass fallowing | 3 | 3 | . | X | 50 |
| Grass fallowing | 4 | 3 | . | X | 57 |
| Grass fallowing | 4 | 2 | . | X | 67 |
| III. Permanent cropping without fallow | | | | | |
| Ley farming | Permanent cropping | – | No fallow | | 100 |
| Tree and shrub cultivation | Permanent cropping | – | No fallow | | 100 |

The economic utilization of this grass fallow for grazing is not, however, practicable for the time being. In the first place nagana is a severe handicap to cattle husbandry in some areas of Africa and secondly, the prices of milk and beef have not yet reached levels where production is profitable. Grass fallow is therefore a measure for maintaining soil fertility which, although technically possible, is uneconomic. The costs of tillage, seed and fertilizers can hardly be offset by the effects of the additional fertility at this stage of development where they are limited to maintaining and increasing soil fertility.

### Transition from fallowing to ley farming

This situation changes the moment the economy reaches a higher stage of development. When sufficient progress has been made in veterinary medicine to provide effective nagana control of tsetse-resistant breeds of cattle, and when national prosperity has increased demand for milk and meat and raised the prices of livestock products, it becomes worthwhile to use the soil preserving grass fallows for grazing cattle. The grass fallows are thus transformed into land under cultivation, and the rate of land utilization now finally reaches 100% (cf. Table 13). The cattle manure can be applied to the fields under crops, and further increases yields. The fields provide the cattle with fodder and the cattle in turn supply the land with manure. Through forage growing and manuring, cultivation of the land and animal husbandry are completely integrated and the symbiosis between these two branches of agriculture that has led to the high productivity of Central European agriculture is established.

### Tree and shrub crops versus field crops

Looking to the more distant future, ley farming, in spite of its advantages, does not appear to be the final stage of development. The growth of forage in the humid tropics is so lush that it would require an enormous livestock population to utilize all that would need to be grown under a ley farming system which aimed to preserve soil fertility. The amount of milk and meat produced would be far above the demand even of a very dense population with considerable purchasing power. It therefore seems unrealistic to regard ley farming as the only possible solution to the problem of soil fertility in the humid tropics. In all biological questions arising in agriculture, we should consult nature, our great master. The natural vegetation in the humid tropics consists of forests. There can be no doubt, therefore, that the fertility of the soil in the humid tropics would be far better preserved by tree and shrub cultivation than by short-term field crops. Tree and shrub plantations come nearest to the natural vegetative cover growing in that climate. Oil and coconut palms, rubber and cocoa trees, tea and coffee shrubs are certainly better suited to preserve the fertility of the soil than are cassava (manioc), taro, yams, sweet potatoes, maize or sorghum millet. In Nigeria soils which have now been under oil palms for 70 years have not lost their natural high fertility. Similarly rubber and cocoa plantations do not seem to have affected soil fertility, provided they have been properly managed.

Because of this some experts hold that when shifting cultivation breaks down, the rural population in the humid tropics should rely to a greater extent on bananas for their basic needs. The starchy variety are well suited to be a basic foodstuff and they ensure the preservation of soil fertility more effectively than short-term field crops. Most other tree and shrub crops, apart from oil and coconut palms, have the grave disadvantage for agricultural countries without economic international trade links that they do not supply basic foodstuffs for the local population. This is why most tropical countries are unable to concentrate on large-scale tree and shrub cultivation to preserve soil fertility. This would require a worldwide trade in food which allowed countries in the humid tropics to exchange the products of tree and shrub plantations for basic foodstuffs, particularly cereals, from the temperate zone. Such a comprehensive pattern of production and trade is, however, imposs-ible to realize. The completely inadequate transport conditions and high transport cost in the humid tropics make the prices the farmer must pay for consumer goods far exceed prices on the world market. For the same reason farmgate selling prices are far below world market selling prices.

In the most distant future, however, such a pattern of international trade in food seems to hold out very promising prospects. It could reduce production costs by making it possible for every climatic zone to adopt those methods of agricultural production which best suited its prevailing climatic conditions and it would help countries in the humid tropics to cope with their soil fertility problems.

## Development Policy Measures to Plan the Evolution of the System

Finally, a few words should be said about the possibilities of planning the evolution of such farming systems through wise development measures so as to lead through increased soil productivity to increased labour productivity. In the course of a country's economic development land will become increasingly scarce so that the cultivation of the soil will have to become increasingly more productive, i.e. more intensive. As an agricultural country changes from being sparsely populated to being densely populated its manpower resources increase although all forms of capital goods continue to be extremely expensive. At this stage of development the necessary intensification of agriculture must be achieved by a higher input of labour. Only when industrialization develops more widely does manpower, as well as land, become scarce and more expensive, although capital goods produced by industry become cheaper. At this stage of development any further intensification must be achieved by increasing the input of capital.

In the course of the process of economic development, then, the productivity of the soil must first be increased. At a later stage, the productivity of labour must be increased in the following triad of intensification stages:

Extensive ————▷ labour intensive ————▷ capital intensive

The stages in the evolution of types of farming which we have described in this chapter and summarized in Table 13 corresponds to this triad of intensification stages. The sequence of the stages is realistic in that as the rise in rates of land utilization and the simultaneous increase in yields per ha of cropped land is initially achieved by means of increased input of labour and later by increased input of capital. At first the transition from uncontrolled to controlled shifting cultivation raises only the ratio of land under cultivation, i.e., it increases only the input of labour. The transition from natural to planted forest fallows requires some capital in the form of plant material although this is raised on the individual farms in advance of the fallow period.

Only when hoeing gives way to ploughing do capital investments in the form of draught animals and equipment increase noticeably. The replacement of forest fallow by leguminous fallow requires capital expenditure for seed and the transition from leguminous to grass fallow also requires capital for fertilizers. Finally, the most advanced forms of land utilization, ley farming and tree and shrub cultivation, require considerable additional capital investment either in productive livestock or a permanent stock of plants.

## Conclusions: the Temporary Nature of the Problem

The problem of soil fertility has occupied man since he first tried to cultivate the soil. The different forms of the problem in each climatic zone determine the way it should be dealt with and the measures to be taken. In the humid tropics the problem is at its most acute. Past experience is not encouraging. It has so far been the fate of the tropics that progress achieved in the fields of medicine and veterinary medicine in the struggle against malaria and nagana has led to an increase in the human and livestock populations. This has been accompanied by a cumulative impoverishment of the soil which has reduced the cropping capacity of the land to an alarming extent.

The problem of soil fertility also varies according to the level of economic development a country has reached. In the humid tropics uncontrolled shifting cultivation in sparsely populated agricultural countries keeps soil fertility in balance, leguminous and grass fallowing in overpopulated agricultural countries gives rise to serious problems. At a later stage in development largely industrialized countries may cope with these by ley farming and tree and shrub cultivation which re-establish the natural balance of fertility.

The realization that the present difficulties are merely of a temporary nature should not merely serve as a reassurance but should above all provide us with an incentive to promote the evolution of agricultural systems with great energy so as to overcome the critical phase in agricultural development as quickly as possible.

## Bibliography

(1957) [Diercke world atlas.] Diercke Weltatlas. Ed.92. Braunschweig, German Federal Republic 220pp.

(n.d.) Symposium on the impact of man on humid tropics vegetation. Goroka, Territory of Papua and New Guinea; UNESCO Science Co-operation Office for South East Asia.

Andreae, B. (1964) [Farm types in agriculture. Origin and change in land use, livestock husbandry and farming systems in Europe and abroad, and new methods for defining them. Methodological part of a textbook on farm management.] Betriebsformen in der Landwirtschaft. Entstehung und Wandlung von Bodennutzungs-, Viehhaltuns- und Betriebssystemen in Europa und Übersee sowie neue Methoden ihrer Abgrenzung. Systematischer Teil einer Agrarbetriebslehre. Stuttgart, German Federal Republic; Eugen Ulmer 426pp.

Andreae, B. (1965) [Land fertility in the tropics. Utilization and maintenance. Farm management considerations for work in developing countries.] Die Bodenfruchtbarkeit in den Tropen. Nutzbarmachung und Erhaltung. Betriebswirtschaftliche Uberlegungen für die Arbeit in Entwicklungsländern. Hamburg, German Federal Republic; Paul Parey 124pp.

Andreae, B. (1966) [Replacing the slash and burn farming system in the humid tropics with more productive methods of farming.] Die Überwindung der Waldbrandwirtschaft in den feuchten Tropen durch produktivere Wirtschaftsformen. *Berichte über Landwirtschaft* 44, 684–704.

Andreae, B. (1967) Measures for the introduction of permanent farming in humid tropical areas. Paper presented at a Joint German Foundation/ECA/FAO Seminar on Problems and Approaches in Planning Agricultural Development, 16 October—7 November, 1967, Addis Ababa, Ethiopia 71–92.

Andreae, B. (1971) Problems of improving the productivity in tropical agriculture. *Economics* 3, 109–127.

Andreae, B. (1974) [The farm sector at the agronomic boundaries of aridity.] Die Farmwirtschaft an den agronomischen Trockengrensen. *Geographische Zeitschrift, Beihefte Erdkundliches Wissen* No.38 67pp.

Andreae, B (1976) [Spatial limits to the food problem.] Räumliche Grenzen des Nahrungs- spielraumes. *Naturwissenschaftliche Rundschau* 29, 393–400.

Andreae, B. (1978a) [Agriculture under local stress.] Ackerbau unter Standortstress. *Berichte über Landwirtschaft* 56 (2/3), 289–307.

Andreae, B. (1978b) [Fallowing in world agriculture.] Brachhaltung in der Weltlandwirtschaft. *Naturwissenschaftliche Rundschau* 31.

Bartlett, H.H. (1956) Fire, primitive agriculture and grazing in the tropics. In: Thomas, W.L. (*Ed*) Man's role in changing the face of the earth. Chicago, Illinois, USA; University of Chicago Press 692–720.

Beguin, H. (1960) [Development of agriculture in Southeast Kasai: essay on agricultural geography and agrarian geography, and possibilities of its practical application.] La mise en valeur agricole du sud-est du Kasai; essai de géographie agricole et de géographie agraire et ses possibilités d'applications pratiques. *Publications, Série Scientifique, Institut National pour l'Etude Agronomique du Congo* 88.

Biebuyck, D. (*Ed*) (1963) African agrarian system. London, UK; Oxford University Press 407pp.

Borchert, G. (1963) [Natural geographical limits of development opportunities in tropical Africa.] Die naturgeographischen Grenzen der Entwicklungsmöglichkeiten im tropischen Afrika. *Die Erde* 94, 313–320.

Bourke, M.R. (1974) A long term rotation trial in New Britain, Papua New Guinea. Ibadan, Nigeria; IITA. [Mimeograph].

Brandt, H. (1971) [The organization of peasant farms under the influence of the development of an industrial town: the case of Jinja, Uganda.] Die Organisation bäuerlicher Betriebe unter dem Einfluss der Entwicklung einer Industriestadt: Der Fall Jinja/Uganda. *Zeitschrift für Ausländische Landwirtschaft, Materialsammlung* No.16 154pp. + lvi.

Brendl, O. (1964) [Agriculture in the Federal Republic of Cameroon in the past and the present, Part 2.] Die Landwirtschaft der Bundesrepublik Kamerun in Vergangenheit und Gegenwart, Teil2. *Bodenkultur* 15, 165–194.

Cabot, J. (1961) [In Chad, the problem of the Koros department of Logone: the example of the Sar plateau.] Au Tchad. le problème des Koros département du Logone: l'exemple du plateau de Sar. *Anales de Géographie* 70, 621–633.

Cleave, J.H. (1974) African farmers. Labour use in the development of smallholder agriculture. New York, USA; Praeger 253pp.

Conklin; C.H. (1957) Hanunóo agriculture. A report on an integral system of shifting cultivation in the Philippines. Rome, Italy; FAO 209pp.

Dumont, R. (1957) Types of rural economy. Studies in world agriculture. London, UK; Methuen and Co. 556pp.

Falkner, F.H. (1938) [The Arid limits of rain fed agriculture in Africa.] Die Trockengrenze des Regenfeldbaues in Afrika. *Petermanns Mitteilungen* 84, 209–214.

Feder, E. (1973) [Agricultural structure and underdevelopment in Latin America.] Agrarstruktur und Unterentwicklung in Lateinamerika. Frankfurt/Main, German Federal Republic; Europäische Verlagsanstalt 307pp.

Finck, A. (1963) [Tropical soils. Introduction to principles of soil science in tropical and sub-tropical agriculture.] Tropische Böden. Einführung in die Bodenkundlichen Grundlagen tropischer und subtropischer Landwirtschaft. Hamburg, German Federal Republic; Paul Parey 188pp.

Flinn, J.C.; Langemann, J. (1976) Experience in growing maize using improved technology in South Eastern Nigeria. Ibadan, Nigeria; IITA. [Mimeograph].

Franke, G. et al. (1967) [Useful crops in the tropics and sub-tropics.] Nutzpflanzen der Tropen und Subtropen, Vols I and II. Leipzig, German Democratic Republic; S.Hirtzel Verlag 324pp.; 421pp.

Freeman, J.D. (1955) Ibadan agriculture. London, UK; HMSO 146pp.

Frey, H.J. (1975) [Empirical studies of labour requirements in small farms in highland areas of Kenya.] Empirische Untersuchungen über Arbeitszeitbedarf in Kleinbetrieben im Hochland Kenias. *Zeitschrift für Ausländische Landwirtschaft* 15, 351–364.

Gaide, N. (1954) [In Chad, transformation of traditional agriculture under the influence of cotton growing.] Au Tchad, les transformations subies par l'agriculture traditionelle sous l'influence de la culture cotonnière. *Agronomie Tropicale, Office de la Recherche Scientifique et Technique Outre-mer* 11, 597–643.

Geuting, H. (1961) [Economic problems and lines of development of agricultural countries in Southeast Asia.] Wirtschaftliche Probleme und Entwicklunglinien südostasiatischer Agrarstaaten. In: Beiträge zur landwirtschaftlichen Betriebslehre. Festgabe zum Geburstag von Professor Dr. H.C. Georg Blohm. Stuttgart, German Federal Republic; Eugen Ulmer, 37–58.

Gourou, P. (1961) The tropical world. Its social and economic conditions and its future status. Translated by E.D.Laborde. London, UK; Longman Ed.3 159pp.

Greenland, D.J.; Herrera, R. (1975) Shifting cultivation and agricultural practices. Ibadan, Nigeria; IITA.

Gullemin, R. (1956) [Evolution of autochthonal agriculture in the savannahs of Oubangui.] Evolution de l'agriculture autochthone dans les savannes de l'Oubangui. *Agronomie Tropicale* 11, (1/3).

Hanrath, J.J. (1964) [Economic problems of arid and semi-arid regions.] De economische problematiek der aride en semi-aride gebieden. *Tijdschrift van het Koninklijk Nederlandsch Aardrijkskundig Genootschap* 81, 172–181.

Harrison Church, R.J. (1961) Problems and development of the dry zone of West Africa (with discussion). *Geographical Journal* 127, 187–204.

Haugwitz, H.W. von; Thorwert, H. (1972) Some experiences with smallholder settlement in Kenya 1963/64 to 1966/67. *Afrika-Studien* No.72 104pp.

Hendrick, F.L. (1960) [Cultivation systems and their evolution in black Africa.] Les systèmes de culture et leur evolution en Afrique noire. Bulletin Institut Agron. Stat. Rech. Gembloa, Horse Serie, Vol.2, p.642 ff.

Heyer, J.; Maita, J.K.; Senga, W.M. (Editors) (1976) Agricultural development in Kenya. Nairobi, Kenya; Oxford University Press 372pp.

Hohnholz, J. (1975) [Agricultural geographical observations in northern Thailand.] Agrargeographische Beobachtungen im Norden Thailands. *Naturwissenschaftliche Rundschau* 28, 311–322.

Jin-Bee, O. (1959) Rural development in tropical areas, with special reference to Malaga. *Journal of Tropical Geography* No.12 222pp.

Johnston, B.F. (1958) The staple food economics of western tropical Africa. Stanford, California, USA; Stanford University Press 305pp.

Jolly, A.L. (1959) Mixed farming in the tropics. *Journal of the Agricultural Society of Trinidad Tobago* 59, 207–210.

Jones, W.O. (1959) Manioc in Africa. Stanford, California, USA; Stanford University Press 315pp.

Kanitkar, N.V. (1960) Dry farming in India. New Delhi, India; Sree Savaswaty Press 470pp. [Ed.2, enlarged with a supplement by S.S.Sirur and D.H.Gokhale].

Kool, R. (1960) Tropical agriculture and economic development. Wageningen, Netherlands; Veenman 151pp.

Lang, H. (1976) Semi-mechanized upland cultivation in African small-scale farms—experiences of the Ivory Coast. *Zeitschrift für Ausländische Landwirtschaft* 15, 220–233.

Manshard, W. (1961) [Fields in forest clearings and strip fields in forests in Africa. Little known elements in the agricultural landscape of Upper Guinea.] Afrikanische Waldhufen- und Waldstreifenfluren. Wenig bekannte Formenelemente der Agrarlandschaft in Oberguinea. *Die Erde* 92, 246–258.

Manshard, W. (1966) [Shifting agriculture and temporary land use for agriculture in the tropics. A comparative review with particular reference to conditions in Africa.] Wanderfeldbau und Landwechselwirtschaft in den Tropen. Eine vergleichende Übersicht unter besonderer Berücksichtigung afrikanischer Verhältnisse. In: Heidelberger Studien zur Kulturgeographie. Festgabe zum 65. Geburstag von Gottfried Pfeifer. Wiesbaden, German Federal Republic; Franz Steiner Verlag 371pp.

Masefield, G.B. (1952) Farming systems and land tenure. *Journal of African Administration* 4 p8 et seq.

Moody, K. (1975) Weeds and shifting cultivation. *PANS* 21 (2).

Mullick, M.A. (1973) [What is the position of the Green Revolution?] Wie steht es um die Grüne Revolution? *Entwicklung und Zusammenarbeit* 4, 10–12.

Nelliat, E.V. et al. (1974) Multi-storaged cropping. *Wild Crops* 26 (6).

Nicholls, L. (1961) Tropical nutrition and dietetics. London, UK.

Niederstucke, H. (1970) [Land use forms in tropical highland areas.] Bodennutzungsformen in tropischen Höhenlagen. *Landwirt im Ausland* 4, 74–76.

Nitz, H.J. (1976) [Land development and change in the cultivated landscape at the boundaries of the world's settled regions.] Landerschliessung und Kulturlandschaftswandel an den Siedlungsgrenzen der Erde. *Göttinger Geographische Abhandlungen* 66, 11–24.

Nye, P.H.; Greenland, D.J. (1960) The soil under shifting cultivation. *Technical Communication, Commonwealth Bureau of Soils* No.51 156pp.

Luwasanmi, J.A.; Adeboyejo, A.F. (1962) Economics of permanent cultivation in the forest regions of Nigeria. *Nigerian Grower and Producer* No.6, Special Issue 4pp.

Pauwels, F.M. (1960) [Agricultural economic research among the Jupaliri.] Landhuishoudkundig onderzoek bij de Jupaliri (Ituri, Oostprovincie, Kongo). Gent, Belgium.

Pereira, H.C. et al. (1958) Water conservation by fallowing in semiarid tropical East Africa. *Empire Journal of Experimental Agriculture* 26, p213 et seq.

Pereira, H.C. et al (1962) Effects of peasant cultivation practices in steep streamsource valleys. *East African Agricultural and Forestry Journal* 27, p104.

Phillips, T.A. (1956) An agricultural note book (with special reference to Nigeria). London, UK; Longmanns 248pp.

Riebel, F.J. (1960) [Agricultural development and situation in Nicaragua.] Entwicklung und Stand der Landwirtschaft in Nicaragua. Dissertation, Rheinische Freidrich-Wilhelms Universität 221pp.

Ringer, K. (1963) [Agricultural systems in tropical Africa. Teaching on agricultural systems. Changes to improve agricultural technology.] Agrarverfassungen im tropischen Afrika. Zur Lehre von der Agrarverfassung. Veränderungen zur Hebung der Agrartechnik. Freiburg im Breisgau, German Federal Republic; Arnold Bergstraesser 236pp.

Ruthenberg, H. (1956) [Problems of the transition from shifting agriculture and semi-permanent agriculture to permanent dry farming in Africa south of the Sahara.] Probleme des Überganges von Wanderfeldbau und semipermanenten Feldbau zum permanenten Trockenfeldbau in Afrika südlich der Sahara. *Agrarwirtschaft* 14, 25–32.

Ruthenberg, H. (1967) [Ways of organizing land use and livestock farming in the tropics and sub-tropics, demonstrated by selected examples.] Organisationsformen der Bodennutzung und Viehhaltung in den Tropen und Subtropen, dargestellt an ausgewählten Beispielen. In: Blanckenberg, P. von and Cremer, H.D. (*Eds*) Handbuch der Landwirtschaft und Ernährung in den Entwicklungsländern, Vol.I. Stuttgart, German Federal Republic; Eugen Ulmer 122–208.

Ruthenberg, H. (1976) Farming systems in the tropics. Oxford, UK; Clarendon Press Ed.2 366pp.

Schiffers, H. (1974) [Droughts in Africa.] Dürren in Afrika. *Forschungsberichte der Afrika-Studienstelle* No.47 242pp.

Schlippe, P. de (1956) Shifting cultivation in Africa—the zande system of agriculture. London, UK; Routledge & Kegan Paul 304pp.

Schlippe, P. de (1958) [Hoe cultivation in primitive slash and burn agriculture, a basic problem of many nations at the threshhold of development.] Hackbau in primitiver Waldbrandwirtschaft, ein grundlegendes Problem vieler entwicklungfähiger Völker. [*Publication*] *Arbeitsgemeinschaft für Rationalisierung des Landes Nordrhein-Westfalen* No.34 38pp.

Scholz, U. (1977) [Permanent dry farming in the humid tropics.] Permanenter Trockenfeldbau in den humiden Tropen. *Giessener Beiträge zur Entwicklungsforschung, Reihe I* 3, 45–58.

Smith, C. (1973) A case study of shifting cultivation and regional development in northern Tanzania. *Zeitschrift für Ausländische Landwirtschaft* 12, 22–39.

Statistisches Bundesamt (1972) [Brief country reports: Chad.] Länderkurzberichte: Tschad. In: Allgemeine Statistik des Auslandes. Stuttgart, German Federal Republic; Statistiches Bundesamt.

Strenge, H. von (1960) [Replacing cotton monoculture in the Sudan.] Überwindung der Baumwoll-Monokultur im Sudan. *Übersee Rundschau* 12(9), 38pp.

Tamsma, R. (1964) [Aspects of arid areas.] Aspecten van aride gebieden. *Tijdschrift van het Koninklijk Nederlandsch Aardrijkskundig Genootschap* 81, 142–171.

Tempany, Sir H.; Grist, D.H. (1958) An introduction to tropical agriculture. London, UK; New York, USA; Toronto, Canada; Longmans 347pp.

Terra, G.J.A. (1958) Farm systems in South-East Asia. *Netherlands Journal of Agricultural Science* 6(3), 157–182.

Trapnell, C.G.; Clothier, J.N. (1957) The soils, vegetation and agricultural systems of North-Western Rhodesia. Report of the ecological survey. Lusaka, Zambia; Government Printer 69pp. Ed.2.

Troll, C. (1975) [Comparative geography of the world's high mountain regions.] Vergleichende Geographie der Hochgebirge der Erde. *Geographische Rundschau* 27, 185–198.

Udo, R.K. (1961) Land and population in Otoro district. *Nigerian Geographical Journal* 4, 3–13.

USA, Department of Agriculture (1963) Farm costs and returns—commercial farms by type, size and location (1960/1963). Agricultural Information Bulletin No.230 79pp.

Vageler, P. (1942) [Fundamental considerations of the problem of humus in the tropics and its practical significance.] Grundsätzliche Betrachtungen zur Frage des Humus in den Tropen und seiner praktischen Bedeutung. *Tropenflanzer* 45 (1/2).

Vink, G.J. (n.d.) [The basis of an Indonesian farm.] De grondslagen van het indonesische Landbouwbedrijf. Wageningen, Netherlands; Veenman & Zonen 204pp.

Wakefield, A.J. (1934) Mixed farming and peasant holdings in Tanganyika territory. *Empire Cotton Growing Review* 11, p.87.

Webster, C.C.; Wilson, P.N. (1969) Agriculture in the tropics. London, UK; Longmans 488pp. Ed.3.

Weischet, W. (1977) [The ecological disadvantage of the tropics.] Die ökologische Benachteiligung der Tropen. Stuttgart, German Federal Republic; B.G. Teubner 127pp.

Welte, E. (1963) [Fertility and production capacity of tropical soils. With special attention to problems of shifting agriculture.] Fruchtbarkeit und Leistung tropischer Böden. Unter besonderer Berücksichtigung des Problems des Wanderackerbaues (Shifting agriculture). *Afrika-Heute, Jahrbuch der Deutschen Afrika-Gesellschaft* 221–230.

Werth, E. (1954) [Digging stick, hoe and plough.] Grabstock, Hacke und Pflug. Ludwigsburg, German Federal Republic; Eugen Ulmer 435pp.

West, O. (1958) Bush encroachment, veld burning and grazing management. *Rhodesia Agricultural Journal* 55, 407.

Westphal, E. (1975) Agricultural systems in Ethiopia. Wageningen, Netherlands; Centre for Agricultural Information and Documentation.

White, G. (1974) Natural hazards. London, UK; Oxford University Press 288pp.

Wilhelmy, H. (1973) Amazonia as a living area and an economic area. *Applied Sciences and Development* 1, 115–135.

Wills, J.B. (Editor) (1962) Agriculture and land use in Ghana. London, UK; Accra, Ghana; New York, USA; Oxford University Press 504pp.

Wills, J.B. (1962) The general pattern of land use. In: Agriculture and land use in Ghana. London, UK; Accra, Ghana; New York, USA; Oxford University Press 201–225.

Wissmann, H. von (1956) On the role of nature and man in changing the face of the dry belt of Asia. In: Thomas, W.L. (*Ed*) Man's role in changing the face of the earth. Chicago, Illinois, USA; University of Chicago Press 278–303.

Wright, R.L. (1972) Some perspectives in environmental research for agricultural land-use planning in developing countries. *Geoforum* 10, 15–33.

Wrigley, G. (1961) Tropical agriculture. The development of production. London, UK; Batsford 291pp.

Wrigley, G. (1971) Tropical agriculture. The development of production. London, UK; Faber 376pp.

Zuckerman, P.S. (1973) Yoruba smallholders farming systems. Ibadan, Nigeria; International Institute of Tropical Agriculture.

# Chapter 6

# Increasing the Productivity
# of Irrigated Arable Farming

## Types of Irrigated Farming

### Economic functions of irrigation in the tropics

Artificial irrigation has the effect of reclaiming land and of economizing in land use. It has the effect of land reclamation in dry areas where it is otherwise impossible to grow crops. It economizes on land in places where, although there is enough rainfall and ground water for agriculture, yields can be increased by irrigation. In either case, irrigation is a measure of intensification, which aims to achieve an increase in gross returns which is higher than the increased expenditure. Irrigation increases intensity in two ways, on the one hand the specific intensity is increased and on the other the intensity of operation. The higher specific intensity is expressed in the additional expenditure on irrigation itself and in the consequent extra outlay on fertilizers, tillage and cultivation for a particular crop.

The profitability of irrigation is not, however, due only to the measurable increases in yields that result from higher specific intensity. Artificial watering affects the entire organization of the farm, increasing its production-intensity. To make the best use of irrigation more demanding crops are cultivated, and crops more capable of intensification with a flatter marginal productivity curve are substituted for those less capable of intensification. Intensive crops that are particularly responsive to irrigation become more prominent in the scheme of cultivation. More field fodder and intermediate and secondary crops are grown and the larger fodder base leads to larger herds of livestock. This produces an increased supply of manure which is welcomed, because any artificial watering uses up humus. As soon as the farmer has secure supplies of the uncertain factor, water, production becomes more regular, ability to withstand crises is greater and it is possible to increase in intensity still further. Thus, as a result of irrigation the entire operation becomes more intensive and productivity per ha is raised throughout. This is why the great increase in population during the 19th century led to a rapid growth of irrigated farming all over the world. The total irrigated area in the world in 1890 was estimated by K. Kaerger as 200 000 sq km, by 1925 K. Sapper put it at about 800 000 sq km. In the mid fifties, K. Schroeder estimated that the total irrigated area in the world had already reached approximately 1.1 million sq km. Table 14 shows area irrigated in 1975 for selected countries in ha, as a percentage of cultivated area, and the increase in irrigated area between 1961/65 and 1975. The importance of irrigation is also illustrated by the FAO statement that although only about 15% of the world's cultivated area is irrigated, it provides about 25% of world food production.

The geographical distribution of irrigation primarily conforms to the climatic conditions, especially the relationship between heat and moisture. Towards the equator this relationship generally changes. There is an increasing surplus of total, annual solar heat available to a crop compared to usable moisture. Other things being equal, therefore, the useful effect of artificial irrigation must increase as one approaches the

Table 14   Irrigation in selected countries, 1961 and 1975.

| | Irrigated areas (ha × 10³) | | Cultivated area [1] ha × 10³ | Irrigated areas 1975 as % of area Cultivated | Irrigated |
|---|---|---|---|---|---|
| | 1961–65 | 1975 | 1975 | 1975 | 1961–65 |
| Europe: | | | | | |
| USSR | 9 618 | 14 500 | 332 207 | 4.4 | 150 |
| Bulgaria | 848 | 1 128 | 4 339 | 26.0 | 133 |
| Portugal | 620 | 625 | 3 620 | 17.3 | 101 |
| Romania | 207 | 1 474 | 10 500 | 14.0 | 710 |
| America: | | | | | |
| Ecuador | 446 | 500 | 4 325 | 11.5 | 112 |
| Puerto Rico | 39 | 39 | 145 | 26.9 | 100 |
| Paraguay | 30 | 55 | 1 000 | 5.5 | 183 |
| Africa: | | | | | |
| Egypt | 2 548 | 2 855 | 2 862 | 100.0 | 112 |
| Algeria | 259 | 320 | 7 050 | 4.5 | 123 |
| Libya | 123 | 135 | 2 544 | 5.3 | 110 |
| Mauretania | 3 | 3 | 1 005 | 0.3 | 100 |
| Ivory Coast | 5 | 25 | 9 120 | 0.3 | 500 |
| Asia: | | | | | |
| Japan | 3 176 | 2 675 | 5 573 | 48.1 | 84 |
| Thailand | 1 729 | 3 149 | 16 580 | 19.0 | 182 |
| Rep. Korea | 682 | 915 | 2 418 | 37.8 | 134 |
| Israel | 142 | 182 | 431 | 42.1 | 128 |
| Developed countries | 27 465 | 31 677 | 416 663 | 7.6 | 115 |
| Developing countries | 71 537 | 90 837 | 670 445 | 13.6 | 127 |
| World | 189 262 | 226 764 | 1 506 149 | 15.0 | 120 |

[1] Cultivated area = arable land + tree and bush crops.
Source: F.A.O. (1977) pp. 45–57.

equator. H. Zörner (1923) therefore distinguished three natural irrigation zones in the northern hemisphere: (1) a zone of 'irrigation in exceptional cases', from the Pole to about lat.55°; (2) a zone of 'supplementary irrigation', from lat.55° to about lat.40° and (3) a zone of 'necessary irrigation', from lat.40° to the equator.

Fig.20, a map drawn by Sapper (1932), shows a more detailed picture of the regions in the world where there are any irrigated areas at all. This map has been interpreted by Fels (1954) among others, to the effect that the irrigated areas conform to the desert and mountain areas of countries which are always hot or have hot summers. Following the mountains, the main band of distribution in the Old World extends in a West-East direction, while in the New World it extends North-South. The most important irrigated areas lie in the sub-tropics and in the tropics away from the equator. Particularly concentrated areas of irrigation developed in the hot regions with strictly periodic rainfall, such as the dry-winter monsoon lands of South Asia, or the dry-summer lands with a Mediterranean climate. In these regions, the natural vegetation dies off during part of the year, not because of cold but because of drought and artificial irrigation makes cultivation possible all the year round. The really dry

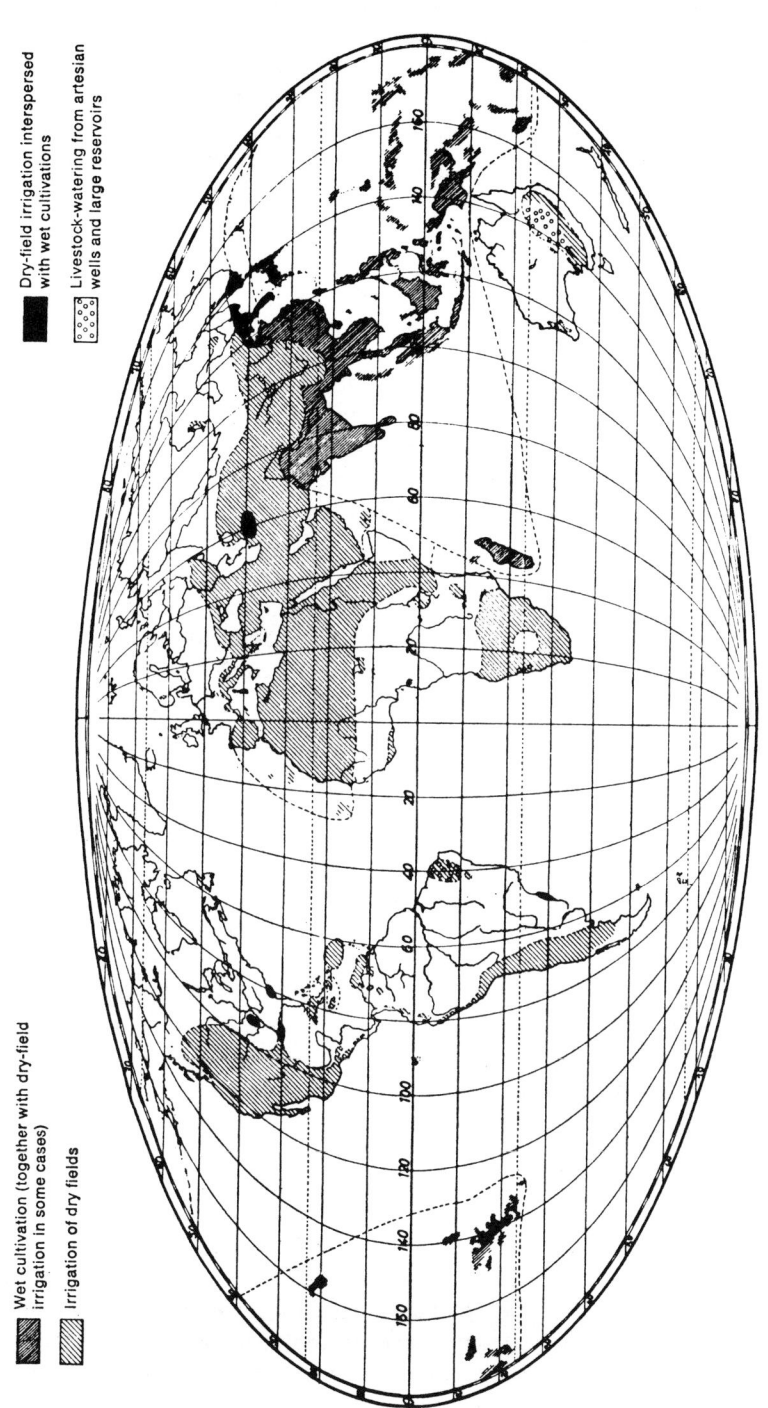

Fig. 20
Geographical distribution of irrigation. (Source: Sapper 1932)

areas (e.g. the Sahara, Egypt, the Sudan) which by their very nature have only desert vegetation are becoming more and more important for irrigation, as dams make it possible to grow large crops. Even in countries with heavy rainfall irrigation has to be considered for commercial crops that require a lot of water. Thus, even in Madagascar, India and Indo-China, China, Korea, Japan and Indonesia tropical wet cultivation, generally of rice, is the rule. Irrigation will also become increasingly important in the tropics of Africa and South America.

Most of the islands in the West Indies are suitable for irrigation. In South America, irrigation is to be found in the entire Andean area, in Patagonia and in the north-east corner of Brazil. In North Africa and East Africa, irrigation is indispensable up to a line connecting Dakar, Sokoto, Lake Chad, Fashoda, Lake Victoria and Mombasa. To this must be added the South African triangle approximately up to the line connecting Benguela and the mouth of the Zambesi, as well as the whole of Madagascar. The part of Asia involved is the southern half divided by a line which runs north of the Aral Sea and Lake Balkhash and reaches the sea to the north of Vladivostock, taking in most of Manchuria. Indonesia and the whole of Oceania, including Hawaii, are included in this irrigated area (Fels 1954 pp.100–101).

The importance of irrigation for different countries may be illustrated by some figures in Table 14. Egypt is the only country in the world whose agriculture depends almost wholly on artificial irrigation; its total cultivated area is the same as its irrigated area. It has only 2.9 million ha of land under cultivation, but its crop acreage is 4.9 million ha, because irrigation makes possible an average of 1.7 crops per year.

## Economic comparison of types of irrigation

The types of artificial irrigation are distinguished according to the nature of the water distribution on the cultivated areas. Strube & Brandes (1972) distinguish the following principal types of irrigation:
1. Furrow-damming irrigation in furrows and ditches (sub-irrigation).
2. The basin or check method of irrigation, from basins or ponds. Both types, (1) and (2), are carried out with still water and can be applied only in terrain with few slopes.
3. Flush irrigation, in which water is supplied to sloping land by being allowed to flow over it in a layer 1–3 cm deep at moderate speed. This uses a great deal of water.
4. Underground irrigation or "reversed drainage".
5. Sprinkler irrigation in a layer of as uniform depth as possible, with uniformly distributed sprinkling. This can be used on any kind of terrain and is very economical with water, compared to surface-irrigation methods.

In furrow-damming irrigation, the water forces its way up to the top soil from underneath, as it does in underground irrigation. In the other types of irrigation the water sinks down from above. Furrow-damming like "reversed drainage" and sprinkler irrigation therefore conserves the crumb structure of the soil. This is endangered by basin irrigation and flush irrigation. On the other hand, furrow irrigation has a number of drawbacks:
1. the loss of useful acreage can be up to 20%;
2. the furrows create difficulties in cultivation;
3. the furrows require continual maintenance;

4. water distribution is uneven and the irrigation effect at the end of the furrows is slight in certain circumstances;
5. erosion damage and leaching can occur,
6. weeds grow in the furrows and
7. there is a danger of the soil being spoiled by salt and of the groundwater rising.

In the basin method of irrigation, the air supply to the soil is cut off. When the water has drained away and the excess water has evaporated, the soil surface left behind is often clogged with mud and encrusted. This further prevents the exchange of gases and thus harms plant growth.

Underground irrigation (sub-irrigation) has some advantages over surface irrigation (1 to 3 above). It carries the water directly to the root area of the plants and thus avoids any water-wasting evaporation and or harm to the mellow soil which is ready for tilling. This method is, however, very expensive. It has the additional drawback that compared with surface irrigation the installations can be more difficult to maintain and repair, and that it is also more difficult to check the irrigation process. Sprinkler irrigation which imitates natural rainfall, saves water, because the watering can be metered much more efficiently than in any other system. It is chiefly suitable for areas where there is not too much evaporation, but it is also important for arid areas.

Table 15   Influence of irrigation on soil fertility.

| Advantages | Drawbacks |
| --- | --- |
| — Weathering is accelerated | — Water erosion, silting up, encrustation |
| — Possible addition of nutrients | — Leaching out of nutrients |
| — On sandy soils, possible enrichment with silt and clay is advantageous | — Depletion of bases |
| — Promotion of plant growth above and below ground, giving more humus production | — On heavy soils, any enrichment with silt and clay is a drawback |
| — Plant growth during the dry period is made possible, reducing wind erosion and plant destruction | — Breakdown of humus is accelerated |
| — The layer of water standing on the soil protects it, like primeval forest cover (against the sun's rays and against too rapid breakdown of humus) | — Danger of salination by chlorides, sulphates, carbonates (calcium chlorosis) and nitrates (impairment of structure) |
| | — Removal of topsoil from parts of the land through levelling of fields for furrow irrigation or irrigation by the basin method |
| etc. | etc. |

Any artificial irrigation influences soil fertility. Table 15 shows the advantages and drawbacks. Examination of these in relation to sprinkler irrigation and surface irrigation soon shows that sprinkler irrigation and underground irrigation are far superior to the surface types of irrigation from the point of view of soil fertility. Sprinkler irrigation requires no levelling of the ground, it uses much less water and it applies it more carefully in imitation of natural rainfall. All of the drawbacks listed are therefore greatly diminished or even avoided altogether, whereas the advantages are largely retained. Sprinkler irrigation also increases the land productivity compared to surface irrigation because there is no loss of useful acreage in ditches or furrows, water is more evenly distributed, seepage losses are lower, and the irrigation effect is greater. Irrigation with sprinkled water is undoubtedly very much more beneficial to soil fertility than irrigation with static or flowing water. The next section therefore, compares sprinkler irrigation with all other forms of irrigation (under the collective term 'surface irrigation') and tries to show where it is likely to be superior.

## Comparative appraisal of irrigation methods

### Topographical conditions

The relative advantages of sprinkler irrigation and surface irrigation are chiefly determined by topographical conditions. Surface irrigation generally requires flat locations or at least graded uniformly inclined slopes (for surface irrigation in the narrower sense). In areas with irregular terrain or shallow soils, preparation of the land for surface irrigation is expensive. With surface irrigation on slopes the water flows away at the end of the field and carries away the soil. Surface irrigation causes erosion damage even with a gradient of only 1:50, whereas sprinkler irrigation causes no such damage even with a gradient of 1:10 (Bandini 1959). Thus if there are no flat sites surface irrigation requires costly grading or terracing merely to prevent erosion. These expensive earthworks can be avoided by using sprinkler irrigation which does not depend upon features of the terrain and does not require any special preparation of the areas to be irrigated. Sprinkler irrigation is, therefore, more suitable for areas with sharp relief and surface irrigation for flat locations. Mesopotamia which is flat has only surface irrigation, while the hilly part of northern Iraq, which has not yet been irrigated, is ideally suited to sprinkler irrigation (see Caesar 1960). So long as only a small proportion of the cultivated land in a country is irrigated, an attempt is made to find flat sites and almost all irrigation is surface irrigation. Later, when irrigated crops also penetrate into hilly territory, where costly grading or terracing would be necessary for surface irrigation, sprinkler irrigation becomes important. Such land has the advantage over flat sites that pumping equipment may not be necessary because the natural gradient produces sufficient nozzle pressure by itself. The more irrigation spreads to uneven terrain, the more are the opportunities for sprinkler irrigation. This fact alone partly explains why sprinkler irrigation has become more important in the course of national economic development and why future irrigated agriculture is likely to conserve soil fertility better.

### Water conditions

Existing water conditions are also a very important factor in the choice of sprinkler or surface irrigation. Account must be taken of the amount and distribution of rainfall and of the amount of water available for irrigation and its cost. To judge this, one must remember that surface irrigation with high capital and maintenance costs has different initial costs to sprinkler irrigation but that the sprinkler system may have greater running costs in certain circumstances. With surface irrigation, fixed costs are relatively higher; with sprinkler irrigation, variable costs. As elsewhere in the economy high fixed costs require regular, frequent and large-scale use of fixed capital equipment so as to spread fixed costs over a large enough volume of production to keep down costs per unit of output. If the extensive preparatory works for surface irrigation are to be carried out (grading the land, building dams, embankments, channel-, ditch- and furrow-systems, bridges, culverts and passages), one must be sure that these structures can be used to increase output. If irrigation is not necessary one year then the high fixed costs of these structures have to be written off. All that is saved is the small amount of wages for start-up and fuel for the pumping plant. If, on the other hand, a sprinkler irrigation installation is not used, much more of the costs can be saved.

Surface irrigation is therefore more suitable for the regular supply of large amounts of water, and sprinkler irrigation more suitable for the occasional supply of small amounts. With surface irrigation, the range of individual applications is to 50–100 mm, depending upon terrain, climate, soil, water supply, type of crop, time and purpose, while for sprinkler irrigation the range is 15–50 mm. Where rainfall is quite inadequate in quantity or distribution for the growth of the crops and large amounts of water have to be supplied regularly, surface irrigation is to be preferred. This method is therefore predominant in all arid countries such as Egypt, the Sudan or Iraq. The high fixed costs of surface irrigation can easily be spread especially in climates where irrigation makes it possible to grow several crops a year from the same field. In a humid climate, on the other hand, irrigation is not normally necessary. It is required only to bridge over occasional periods of drought and then not so much to increase yields, as to save the crop. For this sprinkler irrigation is to be preferred. In the longitudinal valley of California about 90% of irrigation is based on surface watering, whereas in the neighbouring northern state of Oregon, with much more rainfall and damper air, about 90% of irrigation is based on sprinkling. The winter vegetables and citrus fruit crops in eastern Transvaal, where winters are dry, need regular irrigation and are therefore a natural case for surface irrigation. On the other hand, vegetable growing around Cape Town where winters are wet, or sugar-cane cultivation on the humid Natal coast, only require insurance against occasional periods of drought. For these sprinkler irrigation is adequate. In the tropics where rainfall varies sprinkler irrigation is successfully used to bridge over the worse dry periods for crops with a long growing-period or for tree and bush crops. Examples are the cultivation of sugar cane in Uganda or of cacao in Zaire (van Beveren 1959). A significant factor in view of the amount of irrigation water available is that sprinkler irrigation offers higher productivity of water used than surface irrigation. On the one hand with sprinkler irrigation it is possible to control the quantity and timing of the water distributed more exactly and, on the other, the water is more freely and uniformly distributed. Bandini (1959) states that, for Italy, sprinkler irrigation requires only one-half and in exceptional cases only one-fifth of the amount of water needed by other irrigation systems. In Morocco van Beveren points out that cotton grown with sprinkler irrigation requires less than one-third the quantity of water required for surface irrigation. The proportion of water saved depends, of course, upon many circumstances, such as relative humidity, the particular surface-irrigation method compared, etc. Thus, where irrigation water is scarce and expensive and a high marginal productivity of water has to be achieved, sprinkler irrigation is to be preferred. On the other hand, where ample, cheap water is available, surface irrigation with its lower water productivity can be used.

*Soil conditions*

Soil conditions also affect the competition between sprinkler and surface irrigation.

Sprinkler irrigation unlike surface irrigation can be used on a very wide variety of soils. On light soils, surface irrigation results in an uneven distribution of water and high seepage losses, and on heavy soils it sometimes causes flooding. On permeable soils the use of sprinkler irrigation can avoid water losses due to seepage as well as damage due to leaching and salination. Underground irrigation is generally most effective on medium soils. On markedly sandy soils, furrow irrigation is practically impossible. Even on loamy sand soils it is made difficult because water has to be supplied through cement lined pipes or ditches. This makes sprinkler irrigation more competitive. Furrow irrigation is also most effective on medium soils with a stable crumb structure particularly if there is a permeable sub-soil. If the soil becomes heavier and definitely cohesive, sprinkler irrigation again offers various advantages. Heavy soils, particularly in the subtropics, tend to get choked with mud, to harden and to become encrusted because of their lack of humus and it is therefore risky to use surface irrigation on them. Encrustation very much disturbs the growth of plants. Even if the water content of the soil is adequate, irrigation is necessary to keep the surface soft enough for the germinating plant to break through. Basin irrigation is even more disturbing to the soil structure in these conditions. A sprinkler irrigation system which applies smallish amounts of water at fairly short intervals has a better effect, easing plant growth and saving water. In addition sprinkler irrigation partly controls the salt which it continually dissolves and washes into the sub-soil once more (Caesar 1960). Under certain soil conditions in arid areas surface irrigation must be expected to result in spoiling by salt, producing alkali deserts. In humid areas it depletes bases in the soil until it becomes infertile. Such damage is avoided by using sprinkler irrigation.

*Production programme and cultivation methods*

From what has already been said it is clear that the choice between sprinkler and surface irrigation must always also take account of the production programme and cultivation methods. Climate has an indirect influence as well as a direct one upon the relative advantages of the two irrigation systems, because it determines which are the predominant crops.

Salt tolerant crops like date palms are less affected by any saline efflorescence from surface irrigation than salt-sensitive crops for which sprinkler irrigation is relatively more suitable. The date palm tolerates salt so well that the date palm groves around Basra are irrigated by the flood water from the Persian Gulf. Although this water comes from the Euphrates and the Tigris, it is fairly salt in the estuary area. If salt sensitive crops are grown on salty soils or irrigated with salt water, then surface irrigation must be combined with dewatering by drainage. This drainage prevents irrigation water from rising to the surface of the soil and thus prevents saline efflorescences and ensures a downward movement of water. The need for drainage, however, makes surface irrigation much more expensive, and sprinkler irrigation which does not need it becomes relatively more advantageous. Crops which require frequent small applications of water like various kinds of fodder or strawberries are more suitable for sprinkler irrigation. Crops such as rice, which like to stand continually in water, or crops like date palms which need large amounts of water, clearly do better with surface irrigation. The rice fields in the Sacramento Valley of California are put under water only twice a year. The first time is for aerial sowing so that the sinking water draws the seeds into ground. Later they are flooded for the entire remaining growing-period until about four weeks before combine harvesting. Surface irrigation is the only suitable method here. In the Californian Imperial Valley, on the other hand, lettuce is irrigated seven or eight times during its 90-day vegetation period, while lucerne, which in this area gives eight or nine harvests a year, is irrigated twice a month in winter and three times a month in summer. Such crops are also much more suitable for sprinkler irrigation than rice. Crops that form a dense stand of plants covering the whole field, such as grain, fodder crops, vegetables and root crops, derive great benefit from the uniform distribution of water provided by sprinkler irrigation. Tree and bush crops with an open stand and a high water requirement per plant have their needs met better by surface irrigation, which can carry water to each individual plant.

### Irrigation methods in economic development

The extensive network of embankments, ditches and furrows which most forms of surface irrigation require make it difficult to use machinery. Highly mechanized crops and production methods are therefore more likely to be combined with sprinkler irrigation, while with manual working methods, surface-irrigation installations are far less of an impediment. Since manual labour tends to be replaced by machinery in the successive stages of development, sprinkler irrigation also becomes increasingly important.

The relative advantages of the two types of irrigation in different economic conditions of production can best be understood by examining their cost structure two examples of which are shown in Fig. 21. Table 16 which shows costs of irrigation by four different methods under Californian conditions also demonstrates in lines 2–3 that sprinkler irrigation makes the most effective amd therefore most economical use of irrigation water.

The primary consideration in this case is that sprinkler irrigation is a capital-intensive system of irrigation, while surface irrigation is comparatively labour-intensive. Lines 9–10 show that the initial outlay for sprinkler irrigation is 100% expenditure on capital but for surface irrigation 57–67% of initial expenditure is on labour. Lines 16–17 show that for sprinkler irrigation, two thirds of running costs are for materials and only one-third for labour, while for surface irrigation only 18–51% of costs are for materials and 49–82% for labour. At the same time, it must be remembered that in the USA, where wages are very high, relatively capital intensive forms of surface irrigation are used. Labour costs absorb a much higher proportion of running costs of surface irrigation when those surface-irrigation methods which require no wells or pumping equipment are used. Such methods where water reaches the fields through natural flooding or surface irrigation on slopes, or which convey water from the rivers into the channel system by primitive winches or tread-wheels are common in a world context.

The difference in cost structure of sprinkler and surface irrigation as regards the ratio between capital and labour costs also helps to explain their geographical distribution. The industrial countries frequently prefer sprinkler irrigation to surface irrigation, not only because of their natural production conditions, but also because they have to meet the pressure of their high wage levels by a high level of capital investment. In densely-populated developing countries, sprinkler irrigation is practically unknown not only because climatic conditions usually favour surface irrigation, but also because in these countries capital is scarce and labour supplies are

Fig. 21
Capital investment and annual costs of sprinkler and surface irrigation for a 32 ha farm in Montana, USA, 1949. (Source: Heady & Jensen 1954)

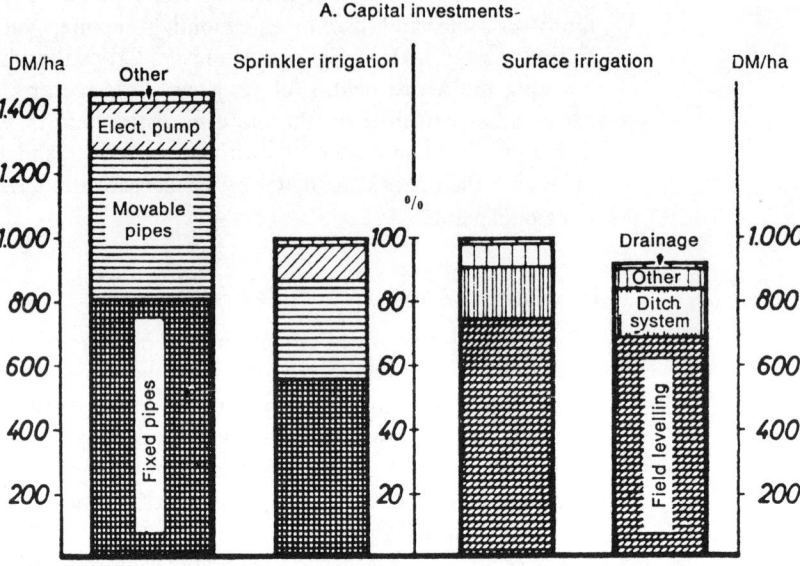

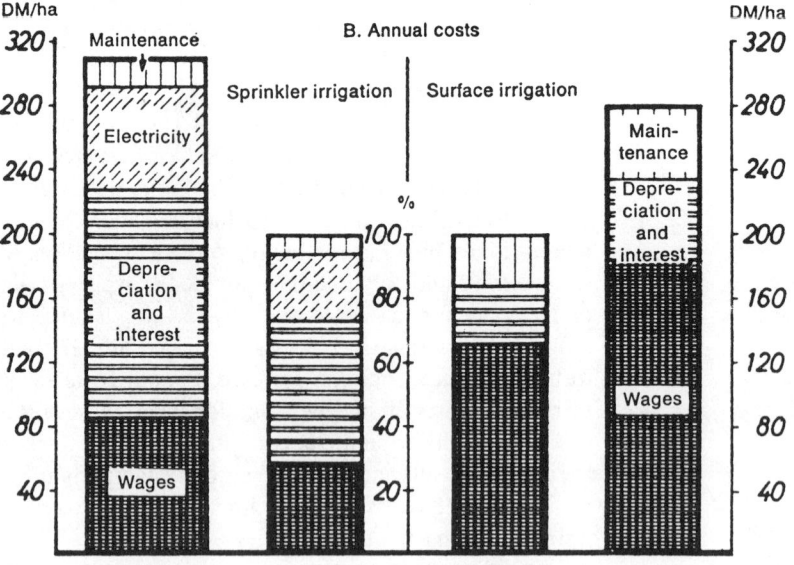

abundant and cheap. The surface-irrigation equipment, which is often still very primitive, is usually built by the farmer himself from simple materials. Tread-wheels or primitive winches are often the only capital equipment, and they are frequently made by the farmer on the farm. In this case the need is to economize on capital, even if this involves a much higher input of labour. In addition agriculture at an early stage of development frequently lacks sufficient technical know-how to use sprinkler-

Table 16   Costs of different methods of irrigation for a 32.4 ha farm in the Californian longitudinal valley.

| | A<br>Sprinkler<br>irrigation | B<br>Surface irrigation | C<br>(furrow-damming) | D |
| --- | --- | --- | --- | --- |
| | | with<br>storage<br>tank | Water pumped directly<br>on to field | |
| Pump capacity, m³/h | 102 | 102 | 102 | 182 |
| Water requirement: | | | | |
| 1. Annual irrigation requirement<br>of the crops in mm | 610 | 610 | 610 | 610 |
| 2. Efficiency of irrigation in % | 80 | 70 | 50 | 70 |
| 3. Total water requirement in mm | 762 | 871 | 1 219 | 817 |
| Investment requirements ($) | | | | |
| 4. Pumping equipment | 2 900 | 2 050 | 1 950 | 2 850 |
| 5. Surface irrigation equipment<br>(invested labour) | – | 4 000 | 4 000 | 4 000 |
| 6. Sprinkler irrigation equipment | 4 800 | – | – | – |
| 7. Water tank | – | 1 000 | – | – |
| 8. Total investment, of which: | 7 700 | 7 050 | 5 950 | 6 850 |
| 9. Capital expenditure in % | 100 | 43 | 33 | 42 |
| 10. Labour expenditure in % | – | 57 | 67 | 58 |
| Running costs in $ p.a.: | | | | |
| 11. Energy costs | 730 | 363 | 342 | 472 |
| 12. Attendance costs | 900 | 742 | 3 456 | 1 388 |
| 13. Fixed costs of pumping equipment | 348 | 245 | 234 | 342 |
| 14. Fixed costs of irrigation equipment | 600 | 375 | 300 | 300 |
| 15. Total running costs, of which: | 2 578 | 1 725 | 4 332 | 2 502 |
| 16. Capital costs % | 65 | 51 | 18 | 40 |
| 17. Labour costs % | 35 | 49 | 82 | 60 |

*Assumptions:* Water level in well = 15.24 m. The nonrecurring levelling costs for surface irrigation are exclusively labour costs (job contractors) and amount to $ 124/ha. For installing the surface irrigation system, the cost is $ 100 p.a. for the whole farm. Pump and sprinkler irrigation equipment depreciated over 10 years, water tank over 20 years. Interest = 5% of current value of all investments. Hourly wage = 90 cents. For surface irrigation, one man must be in constant attendance. Sprinkler irrigation is in 10 instalments, requiring each time 3.1 man-hrs/ha. In addition to attendance costs, there are installation and maintenance costs for the surface irrigation system as labour costs.
Source of basic data: Booker (1952).

irrigation equipment successfully. It is clear from the structure of costs that sprinkler irrigation must gain importance in many countries as their economies develop. This development is characterized by the fact that the production factor labour becomes more expensive, while the production factor capital becomes cheaper. All sectors of agriculture will therefore tend to replace labour with capital. This applies to irrigation and successive stages of labour saving on an irrigated farm might be as follows:
1. Water is supplied by tread-wheels or primitive winches to an open ditch system. Labour input is very high and material costs very low.

2. Water is supplied by fairly cheap low-powered pumps which because of their low rate of water supply require long periods of irrigation, and high wage costs. The open-ditch system (method C in Table 16) is still used.

3. The power of the pumping equipment is increased to shorten irrigation periods and this increases capital expenditure, but saves on wages. This is still an open-ditch system (method D in Table 16).

4. A reservoir is constructed to shorten irrigation times even with low-powered pumps and the open-ditch system (method B).

5. The main feeder ditches are replaced by underground pipelines. This capital expenditure saves considerably on wages because ditching-works are no longer necessary, there is no danger of dams bursting and less supervision is required during surface irrigation. The pipes also prevent loss of water by evaporation, increase the acreage that can be cultivated, make it easier to use machines and avoid the weeds and pests which appear on the edges of ditches.

6. Sprinkler pipes are connected to the underground pipes (see line 5 on Table 16). Laying them requires about the same expenditure on labour as installing the surface-irrigation system in line 5, but all the preparatory work necessary for surface irrigation can now be avoided (grading, furrowing, etc.). The capital costs for sprinkler pipelines and sprinklers are of course considerable and this is probably the most labour-saving, but most costly in capital of the irrigation systems.

Such developments in surface irrigation involving increasing capital expenditure and savings in labour can be observed to-day in many countries which are undergoing rapid economic development. In some cases development progresses as far as the stage of sprinkler irrigation. California, the Republic of South Africa, Italy and Spain are good examples of this process. Progress is hampered by the fact that investment already made in surface-irrigation equipment cannot be withdrawn and this acts as a brake on development. Countries and farms that introduced surface irrigation early, when wages were low and the prices of capital goods high, often continue with surface irrigation for this reason although sprinkler irrigation would be more profitable if a new irrigation system was being installed. This is a correct decision on the basis of marginal costs. Cases therefore occur where, of two entirely similar natural areas with identical cost relationships one has surface irrigation and the other sprinkler irrigation. This is because the first area had been practising irrigation from an early period, while the other area only introduced it after sprinkler irrigation had become more profitable. The change from surface to sprinkler irrigation may also be prevented because of lack of funds for the necessary new investment. The decision is not so much a question of the profitability of the system as the availability of finance.

## An International Comparison of the Economics of Wet Rice Production

Between 1948/52 and 1976 the world's cereal area expanded from 611.5 to 759.4 million ha, or by 24%. Over the same period, the area under rice expanded even more, from 102.6 to 142.2 million ha, or by 38%. The acreage under rice thus increased from 16.8% to 18.7% or the total world acreage of cereals. The importance of rice in meeting the world's cereal requirements becomes even more evident if we take into account that in 1976 rice (paddy) provided 345.4 million tons or no less than 23.4% of total world cereal production [FAO 1976 1977].

**Rice's part in cereal production in various continents**

Rice is grown in all continents, but not to the same extent in each of them (see Table 17). Australasia has 87 000 ha of rice (0.8% of cereal area), Europe 0.6%, North and Central America 1.7% and Africa 6.5%. In South America with 18.8% and particularly in Asia with 37.9%, rice is the major crop. The importance of rice in the overall structure of cereal production is only slightly reflected in the yield ratios. The yield ratios of the four major cereals in six continental areas are shown in Table 18.

From the ecological point of view rice is most closely related to maize. Both are cultivated in the lower latitudes and require high humidity although in this respect rice is considerably more demanding than maize. Both are also the basic food of

Table 17    Rice as part of world cereal production (1976).

| Cereal | World 1961–65 | World 1976 | Europe | N and C America | South America | Asia | Africa | Austral-asia |
|---|---|---|---|---|---|---|---|---|
| | | | Harvested area (ha × 10⁶) | | | | | |
| Wheat | 210.4 | 235.3 | 26.7 | 40.6 | 12.3 | 78.2 | 8.9 | 9.0 |
| Barley | 68.1 | 93.4 | 18.6 | 8.1 | 1.1 | 24.4 | 4.7 | 2.4 |
| Rye | 27.8 | 16.4 | 5.7 | 0.6 | 0.5 | 0.6 | 0 | 0 |
| Oats | 33.3 | 30.2 | 6.1 | 7.6 | 0.6 | 3.3 | 0.4 | 1.1 |
| Maize | 99.4 | 118.1 | 11.7 | 38.3 | 16.5 | 28.5 | 19.7 | 0.1 |
| Millet | 66.9 | 72.8 | 0 | . | 0.2 | 53.2 | 16.3 | 0 |
| Sorghum | 38.5 | 43.9 | 0.2 | 7.8 | 2.4 | 19.0 | 13.9 | 0.5 |
| Rice (paddy) | 124.1 | 142.2 | 0.4 | 1.8 | 7.8 | 127.2 | 4.6 | 0.1 |
| Total cereals | 676.7 | 759.4 | 70.8 | 105.6 | 41.5 | 334.6 | 71.2 | 13.2 |
| Rice as % of cereals | 18.3 | 18.7 | 0.6 | 1.7 | 18.8 | 37.9 | 6.5 | 0.8 |
| | | | Yields in dt/ha | | | | | |
| Wheat | 12.1 | 17.7 | 31.9 | 21.0 | 13.1 | 14.2 | 9.6 | 13.8 |
| Barley | 14.7 | 20.3 | 30.1 | 23.5 | 12.5 | 14.7 | 10.5 | 13.3 |
| Maize | 21.7 | 28.3 | 38.0 | 45.1 | 16.5 | 19.4 | 12.0 | 41.7 |
| Rice (paddy) | 23.6 | 24.3 | 47.6 | 39.7 | 17.5 | 24.6 | 17.4 | 51.3 |
| All cereals | 14.6 | 19.5 | 31.0 | 30.6 | 16.0 | 16.9 | 9.8 | 14.3 |
| Rice as % of average | 162 | 125 | 154 | 130 | 109 | 145 | 178 | 358 |
| | | | Production in t × 10⁶ | | | | | |
| Wheat | 254.4 | 417.5 | 85.2 | 85.4 | 16.1 | 110.7 | 10.8 | 12.4 |
| Barley | 99.7 | 189.7 | 56.0 | 19.0 | 1.3 | 35.8 | 4.9 | 3.2 |
| Maize | 216.1 | 334.0 | 44.5 | 172.8 | 27.3 | 55.1 | 23.7 | 0.4 |
| Rice (paddy) | 253.2 | 345.4 | 1.8 | 7.0 | 13.6 | 312.5 | 8.0 | 0.4 |
| Total cereals | 987.8 | 1477.3 | 219.5 | 322.6 | 66.3 | 566.1 | 69.7 | 18.8 |
| Rice as % of total | 25.6 | 23.3 | 0.8 | 2.2 | 20.0 | 55.3 | 11.5 | 21.2 |

Source: FAO, 1977 pp. 89 ff.

Table 18   Comparative yields and area of rice harvested in six continental areas (1976).

| Area | Yield ratios | | | | | | | % rice in cereal area |
|------|-----|---|-------|---|-------|---|--------|------|
|      | Rice [1] | : | Maize | : | Wheat | : | Barley | |
| Europe | 100 | : | 80 | : | 67 | : | 63 | 0.6 |
| North and Central America | 100 | : | 114 | : | 53 | : | 59 | 1.7 |
| South America | 100 | : | 94 | : | 75 | : | 71 | 18.8 |
| Asia | 100 | : | 79 | : | 58 | : | 60 | 37.9 |
| Africa | 100 | : | 69 | : | 55 | : | 60 | 6.5 |
| Australasia | 100 | : | 81 | : | 27 | : | 26 | 0.8 |

1) Paddy
Source: FAO (1977)

populations which still have rather low purchasing power. In North and Central America the yield ratio is 100:114 in favour of maize which, since it also requires less outlay, dominates the field. Rice has thus been able to capture only 1.7% of the cereal acreage and is really only grown widely in two areas in the United States; Arkansas and the Sacramento Valley, California, which has permanent irrigated rice-fields. In South America the rice/maize yield ratio (100:94) shifts far more towards rice which thus accounts for 18.8% of the cereal acreage, particularly since in certain areas the wet tropical climate is favourable to its cultivation. On the yield basis rice gains an even greater dominance over maize in Asia where the ratio is 100:79 and the area under cultivation is 37.9% of the total cereal acreage. In over-populated agricultural countries rice's higher yield per ha becomes a much more important factor than the lower cost of growing maize. The per-caput consumption of rice is 88 kg a year in India, 147 kg in Japan, and up to 200 kg in Malaysia while in most European countries it drops to a mere 0.5 kg (Krug 1957 p.39 et seq.). In a large part of Asia rice provides up to 80% of all calories and 75% of all protein.

The yield ratio between rice and millet is also very small in North and Central America, but millet is a less important crop since maize produces better yields. On the other hand, in Africa rice accounts for only 6.5% of the cereal acreage, compared to 27.7% for maize and 22.9% for millet. The yield ratio, however, is exactly the opposite: 100:56:34. Maize is predominant in America and millet in Africa. This can largely be ascribed to the dry climate of large parts of Africa, the higher level of intensity of American agriculture, and the population's traditional eating habits. The most important rice-growing continent, however, is Asia. The fact that here the area under millet is 42% of that under rice even though the yield of millet per ha is only a quarter of that of rice, is due to the different ecological requirements of the two cereals. Millet is grown only in fields which are unsuitable for rice, interspersed with smaller maize fields in places unsuitable for growing rice, but still too good for millet.

The ratio between rice and wheat yields provides hardly any information on the reasons for the ratio between acreages because rice prefers the lower latitudes and wheat the higher latitudes. However, rice is grown up to 45° N and 40° S. It should be emphasized that of these two major cereals, rice has a markedly higher yield than wheat in all parts of the world. Thus, while the world area under rice is 42% less than the area under wheat, in 1976 the volume of rice produced was only 18% less than wheat. It is fortunate that the overpopulated developing countries in the wet tropics can grow the highest yielding cereal, rice, while the industrial countries in the temperate zones can favour wheat which, although lower yielding, is better suited to their requirements.

The cultivation of rice is possible over a very wide range of economic conditions. It can be found in sparsely populated and over-populated developing countries as well as in the industrialized countries of Europe and America. Rice is also distinguished by a large ecological variability since it is equally as productive in tropical rainy and savanna climates and in semi-arid and semi-tropical regions as it is in the Po Valley and the Rhone delta. Rice can be grown in hilly regions and at sea-level, on rich and poor soils, in irrigated and dry fields. If rice-growing is possible under such a wide range of economic and natural conditions, it is only because very different production methods make it highly adaptable. Table 19 shows this very well.

## Rice growing in major producing countries

Table 19 lists 13 major rice-producing countries classified by their stage of socio-economic development, into the two basic categories of agricultural and industrial countries. The criterion used is the per caput income in US dollars. In the agricultural countries listed it only rarely and slightly exceeds $500 while it amounts to over $2700 in the industrialized countries. Population density is also a significant factor affecting acreage and methods of rice production in under-developed agricultural countries with a low level of food supply. The agricultural countries have therefore been divided into one group with a low population density of up to 80 inhabitants per km² and another with a high population density of 140 inhabitants and more. The reference unit of "km²" is obviously somewhat dubious as 1 km² of alluvial soil is rather different from 1 km² in a mountain or desert area. The process of industrialization is indicated by the declining percentage of the total labour force employed in farming. This industrialization process is a primary factor in the socio-economic development which leads to increased income per caput.

In addition to these basic socio-economic data agricultural production conditions in the various countries significantly affect rice production. The column "climate" in Table 19 shows that virtually all the sparsely populated agricultural countries are in permanently or seasonally wet tropical climates. The best biophysiological conditions for growing rice are found more in the seasonally wet tropical climate or, at least in the borderline between the two. Here, rice's need for high humidity and high temperatures is easily satisfied throughout the year so that, with a vegetation period of 3–6 months, two or three crops a year can sometimes be harvested from the same field. In densely populated agricultural countries rice growing also spreads to less favourable climates because even here, rice generally provides the most nutrients per ha of all cereals, and these countries must, primarily aim at high

Table 19    Rice growing in selected countries: conditions and results.

| Country | Basic socio-economic data | | | Agricultural production conditions | | | Rice production (paddy) | | | | |
|---|---|---|---|---|---|---|---|---|---|---|---|
| | Population per km² (1975) | Percent employed in agriculture (1976) | GDP per caput ($, 1975) | Predominant climate for rice-growing[1] | Tractors per 1 000 ha[7] (1975) | kg N per ha[7] (1974) | Area under cultivation | | | Rice yields in dt/ha (1976) | Wholesale price per dt milled rice ($, 1976) |
| | | | | | | | in 1 000 ha (1976) | % total[7] (1975) | % irrigated[7] | | |
| **A. Sparsely populated developing countries** | | | | | | | | | | | |
| Brazil | 13 | 41.2 | 774[6] | Af; Aw | 7.14 | 10.6 | 6 588 | 18.3 | 693 | 14.5 | 27.9 |
| Indonesia | 20 | 62.0 | 126[5] | Af | 0.81 | 21.6 | 8 800 | 50.4 | 228 | 26.1 | 33.0 |
| Madagascar | 43 | 85.9 | 133[4] | Af; Aw | 0.82 | 1.1 | 1 050 | 35.0 | 234 | 17.3 | . |
| Burma | 46 | 54.8 | 101[6] | Af; Aw | 0.34[2] | 3.7 | 5 180 | 49.8 | 513 | 18.2 | . |
| Thailand | 81 | 77.2 | 342 | Aw | 1.16 | 4.8 | 8 200 | 49.5 | 260 | 18.2 | 18.0 |
| **B. Densely populated developing countries** | | | | | | | | | | | |
| Egypt[2] | 37[3] | 52.0 | 260[5] | Bw | 7.50 | 35.0 | 484 | 16.9 | 16.9 | 52.3 | 12.9 |
| Pakistan | 87 | 55.5 | 368 | Af; Bs; Bw | 1.95 | 18.5 | 1 699 | 87.0 | 119 | 23.2 | . |
| Philippines | 142 | 48.9 | 368 | Af | 0.80 | 22.5 | 3 562 | 45.0 | 256 | 18.1 | 23.4 |
| India | 182 | 66.0 | 137[6] | Af; Aw; Bs; Ca | 1.29 | 10.6 | 38 600 | 23.1 | 119 | 18.3 | 16.7 |
| Korea Rep. | 352 | 43.6 | 551 | Ca; Da | 0.19 | . | 1 215 | 50.0 | 132 | 59.7 | 50.6 |
| **C. Developed (industrialized) countries** | | | | | | | | | | | |
| USA | 23 | 2.6 | 7 087 | Cs; Ca | 19.70 | 37.3 | 1 012 | 0.48 | 6.2 | 52.4 | 28.2 |
| Italy | 185 | 14.0 | 2 704[6] | C | 66.50 | 39.0 | 180 | 0.15 | 5.0 | 54.2 | . |
| Japan | 298 | 14.1 | 4 133[6] | C; Da | 63.00 | 123.9 | 2 779 | 50.0 | 104 | 55.0 | . |

[1] A = Tropical rainy climate: Af = wet, Aw = alternating wet: B = dry: Bs = semi-arid, Bw = arid;
    C = humid temperate climate: Cs = semi-tropical, dry in summer, Ca = semi-tropical, warm in summer;
    D = humid, moderately cool climate: Da = continental, warm in summer.
[2] Cultivated area producing more than one crop a year is doubled.
[3] Most of the area cannot be used for agriculture (desert).
[4] 1970.
[5] 1975.
[6] 1974.
[7] Arable land and permanent crops.
Sources: F.A.O. (1976 and 1977). U.N.O. (1977).

yields from their land. While the semi-arid rice growing areas in India or Pakistan are mostly situated in regions flooded by seasonal rains, they can produce only one crop a year because temperatures are too low and the water supply is mostly inadequate during the dry winter season. In semitropical regions with warm summers, such as Korea or Taiwan, the water supply is ample so that plants requiring less warmth, such as wheat, rape, forage, vegetables or sweet potatoes, can be grown in the winter. Rice growing during the summer is increasingly handicapped by the longer days and falling temperatures as we move towards higher latitudes.

This is particularly the case for the industrialized countries which are normally in higher latitudes. Caesar (1968 p.393) however, points out that the disadvantages of the circadian light stimulation rhythm and the lower temperatures for growing rice in places like southern Europe are more than compensated by the advantages which arise from longer exposure to light. These include prolonged assimilation time per day and possibly also a longer vegetation period as result of the light stimulation rhythm and increased storage of substances during the nights which are cooler than those in the tropical regions. These effects do not, however, provide a full explanation of the particularly high rice yields in southern Spain, northern Italy and Japan. These should rather be ascribed to the capital intensive methods of production used in industrialized countries and especially to the high expenditure on plant protection and on mineral fertilizers.

The next two columns in Table 19 show tractor-ownership to represent the use of labour-saving equipment, and use of nitrogen to represent land-saving materials. In theory we should expect (1) that sparsely populated agricultural countries should necessarily prefer labour-saving capital goods because their aim is high labour productivity; (2) that densely populated agricultural countries would be obliged to prefer capital goods which increase yields and save land because land productivity has priority for them and (3) that industrialized countries would employ both types of capital goods equally and to a considerably greater extent than the developing countries, because with their high land and labour costs they are faced with the difficult task of combining high land productivity with high labour productivity.

As regards the use of tractors hardly any difference can be found between densely populated developing countries because, generally speaking work organization in farming has not yet progressed as far as mechanization. Only industrialized countries show any marked increase in tractor utilization. This is less in the USSR, despite its large agricultural units because socio-economic development is less advanced, while in Italy, despite its small units, it is greater because of the high level of socio-economic development.

Nitrogen utilization rises more rapidly in the densely populated agricultural countries than in the sparsely populated ones. This is particularly true in the two countries with the highest population densities, Korea and Taiwan both of which attain rice yields which come close to those of the industrialized countries. The potentials inherent in the use of mineral fertilizers which are especially important for the overpopulated developing nations of south-east Asia are still largely unknown and unused. Most wet rice fields receive no mineral fertilizer at all, although interpretation of the numerous fertilizer experiments carried out by the FAO (1966) produced net returns (increase in the gross returns less fertilizing costs) of $4.6/ha for the first 30 kg/ha of pure nitrogen in South Vietnam, $7.9/ha in Thailand, $18.5/ha in Ceylon, $30.5/ha in India, and as high as $47.4/ha in Bangladesh. Only in Burma was this type of fertilizer application found useless (G. Kemmler 1968, p.16). The extent to which mineral fertilizers can be used is determined largely by price relationships. In 1967 the producer price of 1 kg rice could cover the retail price of only 0.19 kg of nitrogen in India, 0.23 kg in Thailand and Taiwan, 0.30 kg in Philippines, and no less than 1.08 kg of pure nitrogen in industrialized Japan.

Examination of rice acreage as a percentage of the total and irrigated areas makes it possible to draw some cautious conclusions on the proportions of wet and dry rice growing, and on the number of crops grown per year. From the fact that Burma grows rice on 50% of the total area and 513% of the irrigated area it is safe to conclude that two or three crops a year are frequently obtained and that a considerable proportion of the rice is grown in dry areas. This second fact appears to be an important reason for the low average yield.

In Guyana, the "Blue Belle" variety can provide three or four crops a year each of 3000 to 4000 kg/ha. With three crops, the annual yield is around 12000 kg/ha. (Feistritzer 1969 p.40). The statistics for Sri Lanka suggest a considerable proportion of dry farming which accounts for 53% of the acreage and, depending on the weather, provides only 20–40% of the yield, (K. Caesar 1960 p.393). In Taiwan, where rice is far the most widespread crop and the wet system is the predominant method the fact that the rice acreage is 42% larger than the irrigated areas, is explained by the possibility of two crops a year in some regions. China, Korea, and Egypt grow only irrigated rice, while the dry system accounts for 3% in Japan, for 10% in Malaya, for 15% in Java and Madura, and for 20% in the Philippines. In contrast to south-east Asia, dry farming of rice is the more common system in Latin America.

### Rice production methods in agricultural and industrialized countries

Farm production methods always aim to obtain a combination of minimum costs. The farmer tries to combine his working resources in such a way that the cost of any given production process is minimized. The three production factors—land, labour and capital in the form of an extremely wide range of goods—must be employed in a quite specific ratio which is the minimum cost combination. The particular ratio depends on the cost of the three factors, i.e. rent, wages, and interest.

The use of each production factor is subject to the law of diminishing returns so that if capital is increased while the land and labour remain unchanged, the returns on each additional injection of capital fall progressively. The capital employed should be increased only up to the point where marginal returns and marginal expenditure are in balance, or in other words to the point where the cost of the last unit of capital employed is still covered by a corresponding improvement in performance. Exactly the same rule applies if the expenditure on land is increased while the expenditure on labour and capital remains unchanged, or if the expenditure on labour is increased while the expenditure on land and capital remains unchanged. The marginal productivity of any production factor always drops in proportion to the increased employment of that factor as compared with the others, so that it is always necessary to extend the use of any production factor to the point where its marginal productivity equals its cost. The minimum cost combination is attained if the marginal returns of all three production factors are equal to their marginal costs. The consequence of this is that comparatively expensive production factors which call for high productivity must be used sparingly while the cheapest production factor with its low costs, and low marginal productivity despite diminishing returns can be given the primary role in the production process.

As we have seen, rice can be grown under an extremely wide range of technical production methods, which allow it a very wide economic latitude. Particularly extreme differences in the minimum cost combination are bound to occur if production methods in two countries at opposite poles of economic development are compared. For this purpose, rice growing in an industrialized country, USA, is compared with rice growing in an agricultural country, Thailand. (See Table 3 in Chapter 2.)

## Rice growing: monoculture or crop-rotation

The way in which rice is incorporated in a farming system and particularly whether monoculture or crop rotation is more suitable, depends largely upon the local cropping system for rice. Rice growing systems can be divided into five groups (see Franke et al 1967 p.192 et seq.). These are (1) dry or hill rice cropping; (2) half-dry cropping; (3) wet cropping; (4) swamp cropping; and (5) aquatic or sprout cropping.

Dry or hill rice is cultivated in the same way as other cereals in hilly regions or other areas where irrigation is not possible. In the tropics this dry cropping is usually limited to areas about 800 m above sea-level, with 800-1200 mm rainfall, and an average day temperature of 24°C for at least 4 months. However, in Peru rice thrives in areas up to 1100 m above sea-level, in the Philippines up to 2000 m and in the upper valleys of the Himalayas even up to 3000 m (Krug 1957 p.39 et seq.). Like the swamp and aquatic systems found in flooded regions the hill farming system of growing rice is of minor importance.

Swamp rice is grown on rivers in flood, as for example on the W. Coast of Madagascar. The seedlings are transplanted when the floods have dropped to a low enough level. Aquatic rice or ratoons are sown at the end of the dry season into dried-out lakes and lowlands in areas like Bengal. A growth of 50 cm is needed before the floods begin in the rainy season. Since rice can grow only 5–10 cm a day, the water must not rise too quickly. This cropping system involves a high risk and the yield is rarely more than 1300–1500 kg/ha. The most important and, in Asia, most typical system is wet cropping. From sowing until shortly before harvest the water requirements of the rice are constantly supplied by artificial irrigation. This requires reservoirs, terraces, and embankments. Since widespread flat areas are rare, the irrigated system usually consists of very small separate fields. This and the existence of permanent 40–60 cm high embankments form serious obstacles to mechanization. Fortunately the small farmers in overpopulated agricultural countries do not need mechanization.

The swamp, aquatic and wet cropping systems can be grouped together as water-rice systems in contrast to the dry-rice systems. The semi-dry system falls between them. Its special characteristic is the preparation and sowing of seed beds which are dry. Subsequently the crop is periodically and frequently irrigated. This system is used when water supplies are insufficient for constant irrigation or when mechanized tilling and sowing make earlier flooding impossible.

### Rice in monoculture

Some of these cropping systems, i.e. those in the wet-rice group, automatically require or at least suggest monoculture.
(1) During the rainy season in the Asiatic tropics, extensive areas such as the Bangkok Plain in Thailand or the river basins in Bengal are flooded. Rice constitutes the only possible crop in such areas and all attempts to find suitable alternative crops have so far failed (Piekenbrock 1958 p.24).
(2) In parts of the Sacramento Valley in California, the soil is so heavy and damp that it provides a perfect site for growing rice. The choice lies not between rice and/or other field crops but between rice and wasteland. The result is rice is a monoculture interspersed with fallow areas.

(3) Small farms in over-populated developing countries concentrate on growing rice to supply their basic food needs. In the wet tropics with ample water they make use of the possibility of obtaining two or three crops of rice a year which produces its highest yield in those regions. They adapt themselves solely to a rice diet because rice like sugar-cane, maize, and tuberous roots is one of the crops with a particularly high calorie content (Finck 1970 p.48).

(4) The high cost of terracing and building the reservoirs and embankments needed for wet cropping can be borne only if the fields are in permanent use. Where no other crop is equally satisfactory nor as profitable to irrigate, rice monoculture becomes the obvious answer. On occasion it provides unusual by-products if the rice fields are re-flooded after the harvest. In Arkansas it lures wild-duck while in Madagascar and Java it allows young fish to be farmed.

(5) In some parts of Asia, certain local conditions in the cultivated areas make it impossible to introduce alternative crops. Thus, areas with frequent heavy rainfalls, low-lying, badly-drained basins, low delta regions near the coast, or terraced slopes with a constant flow of water are fit only for rice which suits excessively moist and flooded soils (Franke et al 1967 p.188).

Fig. 22
Rice monoculture with varying degrees of land utilization

| A | B | C | D | E | F |
|---|---|---|---|---|---|
| Short ← | | Vegetation period | | → Long | |
| | | ← Increasing water supplies | | → | |
| 1.[1] Fallow<br>2. Rice | 1. Fallow<br>2. Rice<br>3. Rice | 1. Fallow<br>2. Rice<br>3. Rice<br>4. Rice | 1. Rice<br>2. Rice<br>3. Rice<br>etc. | 1. a) Rice<br>   b) Rice<br><br>2. a) Rice<br>   b) Rice<br>   etc. | 1. a) Rice<br>   b) Rice<br>   c) Rice<br><br>2. a) Rice<br>   b) Rice<br>   c) Rice<br>   etc. |
| % of land utilisation | | | | | |
| 50 | 67 | 75 | 100 | 200 | 300 |
| Examples of countries | | | | | |
| California<br>(Sacramento Valley) | | | Burma,<br>Thailand | South Japan,<br>Java,<br>Puerto Rico | Formosa,<br>Guyana,<br>North Vietnam |

[1] 1, 2. etc. = year;  a), b) etc. = consecutive crops in the same year.

Essentially the prime reason for variations in when rice is grown as a monoculture is the intensity of land use, which in turn depends very much on the length of the vegetation period, the water supply, and the quality of the soil. (see Table 17). However, the need to adopt monoculture need not involve any major disadvantages in growing rice. Rice like rye, cotton, maize and sugar-cane is self-fertile so that even under permanent cultivation, yields rarely drop further after a minimum level of soil fertility has been reached. This can be seen in areas in lower Burma where rice has been grown as a monoculture since the turn of the century. (Franke et al 1967 p.188). Some of the famous terraced slopes in the Philippines have been producing rice for over 2000 years without a break (Ruthenberg 1967 p.161). In other regions, particularly in the semi-tropics, initially satisfactory yields from rice on virgin soils gradually fall off under monoculture because of disease, depletion of nutrients and the spread of weeds. To combat this Uruguay, Texas, and Louisiana, among other areas, are trying to develop suitable crop rotations. Arkansas has already done so.

*Geographical comparisons of crop rotation with rice*

Table 20 summarizes 22 types of crop rotation which include rice. It is not possible to see a clear relationship between the type of crop rotation and the climatic zone or the stage of economic development. The following observations can, however, be made:

(1) Dry rice occurs more in crop rotations because the rotation is not prevented by the desire to make best use of irrigation.

(2) In places like Costa Rica or West Africa with shifting cultivation rice has to form part of a rotation, usually because shifting the fields makes irrigation impossible and only hill rice can be grown. In Malaysia where rice is grown in consecutive years with a shifting cultivation system the yields drop to around 65% as early as the 2nd year after the woodland has been cleared for arable cropping, and to around 46% in the 3rd year. In the Congo the 2nd year of cultivation produces a crop of only about 24% of that in the first year after clearing (Nye & Greenland 1960).

(3) Crop rotations producing more than one harvest a year are particularly common in the humid tropics with constant warm temperatures and, of course, primarily in the densely populated agricultural countries where maximum land utilization is essential. In places where a seasonal vegetative rest is not necessary it is possible to have two or three rice crops a year (North Vietnam, Guyana, Puerto Rico) or rice alternating with another crop in the same calendar year (Java, India, Madagascar).

(4) If a warm season alternates with a cooler one, as in semi-arid regions or warmer semi-tropics, the advantages of two crops in a single year can be gained by growing rice, a tropical plant, in the summer and a temperate plant which requires less warmth, in winter. In the lowlands, moving away from the equator, winter crops like sweet potatoes and soya give way first to those like sorghum and legumes, then to vegetables and forage, and finally to wheat, rape and similar crops. Unlike countries in the humid tropics, Chile, Egypt, Taiwan and Japan (see Table 21) can only grow more than one crop a year if they practise crop rotation.

(5) If the winters are still colder with a vegetative rest, as in Northern California or the cold agricultural zone of Japan, rice again increasingly reverts to monoculture but with only one crop a year.

(6) As the socio-economic situation of a region improves, rice monoculture becomes less common and is replaced by crop rotations which include rice. Rice, the original subsistence crop is now supplemented by various cash crops or the integration of crop husbandry and cattle involves the production of forage which grows well on wet fields with irrigation and competes with rice. Finally biological measures to increase yields and combat diseases, pests, and weeds also encourage crop rotation.

Table 20   Rice in crop rotations as conditioned by climate and stage of socioeconomic development.

| Climate | Sparsely populated agricultural countries | Densely populated agricultural countries | Industrialized countries |
|---|---|---|---|
| Humid tropical | *North Vietnam* (1)<br>The most intensive wet rice cultivation, 3 crops a year from the same field, total yield. 8 000–11 000 kg/ha.<br>*Costa Rica*<br>Hill rice as part of shifting cultivation with maize, beans and manioc | *Java*<br>Rice in rotation with maize, soya, ground-nuts, sugar-cane, sorghum, sesame, manioc or tobacco | *Puerto Rico*<br>2 rice crops a year from the same field:<br>or<br>1–3: grass/clover mixed<br>4–5: rice |
| Tropical with varying humidity | *Madagascar*<br>Rainy season: rice<br>Dry season: soya beans for fodder<br>*West Africa*<br>Many years of bush, 1st year hill rice following years: beans, manioc etc. | *India* (1)<br>Annually: Rice, sorghum or eleusine<br>or<br>1: sugar-cane<br>2: vegetables<br>3: rice<br>or<br>1–4: rice<br>5: tobacco, cotton, betel, groundnuts or sorghum | *Australia*<br>1–5: grass/clover mixed<br>6–7: rice<br>8: oats or wheat |
| Semi-arid and arid | *North Chile*<br>a) winter: wheat<br>b) summer: rice | *Egypt*<br>a) winter: cereals, legumes, forage crops, or vegetables<br>b) summer: rice | *USSR (Kuban oblast)* (2)<br>1–2: grass/clover mixed<br>3–4: rice<br>5: vegetables, fodder roots<br>6–7: rice |
| Semi-tropical, dry in summer | *Turkey*<br>1–2: rice<br>3: wheat<br>or<br>1: rice<br>2: wheat | *Morocco*<br>1: wheat-grass/clover mixed<br>2: grass/clover mixed<br>3–4: rice | *USA* (California)<br>Rice-monoculture<br>or<br>1: fallow<br>2–3: rice<br>or<br>1: fallow<br>2: barley<br>3–5: rice<br>or<br>1: fallow<br>2: rice |
| Semi-tropical, warm in summer | *Uruguay*<br>Rice-monoculture (with falling yields) | *Taiwan* (3)<br>In the same year:<br>a) sweet potatoes<br>b) rice<br>c) vegetables<br>d) rice<br>or<br>a) rice (March-July)<br>b) rice (August–Nov.)<br>c) wheat (Dec.–Feb.) | *Italy (Lombardy)*<br>1–2: grass/clover mixed<br>3–7: rice<br>8: wheat<br>or<br>1–2: grass/clover mixed<br>3: maize<br>4–7: rice<br>8: wheat<br>*South Japan*<br>1a) rape<br>  b) rice<br>2a) range grass<br>  b) rice |

Sources: (1) Franke 1967 p. 189ff; (2) Konnecke 1967 p. 305; (3) Ruthenberg 1967 p. 163.

Table 21   Changing crop rotations due to longer vegetation periods in Japan.

| Rotation A | Rotation B | Rotation C | Rotation D | Rotation E |
|---|---|---|---|---|
| Cold zone | Cool agricultural zones | | Warm agricultural zones | |
| North, 42°←——————————————— 1 600 km. ————————————→ South, 30° | | | | |
| 180–240 veg. days | 240–260 vegetation  days | | over 260 vegetation  days | |
| Year 1–2: red clover Year 3: green maize Year 4–6: rice | Year 1: summer: rice winter: root vegetables Year 2: summer: rice | summer: rice winter: cereals | spring: root vegetables autumn: rice | winter: range grass spring: rice autumn: rice |
| Number of main crops a year | | | | |
| 1 | 1.5 | 2 | 2 | 3 |

Source: Tsuzuki (1963) p. 833 et seq.

*Crop rotations including rice in the course of economic development*

The large rice farms in Arkansas may be taken as an example of the development from monoculture to crop rotation. Pure monoculture, like that used for cotton, maize and sugar-cane, is only possible for rice in this area if combined with fallow. There is a strong tendency for grass and water-weeds to develop, lime accumulates in the topsoil because of the irrigation water, rice diseases and pests increase, and the humus is depleted by the high demands from the rice crop. Since the 1930 world depression, however, rising land values have made fallow areas uneconomic. Development here skipped the stage of diversified production and passed directly from monoculture to specialization. This was firstly because of the high cost of the irrigation and machinery needed for growing rice and secondly because socio-economic integration had advanced to a stage where the conditions for diversification were no longer present. As a result in summer the fallow land and the areas which had become unsuitable for growing rice for the reasons listed above were sown with crops selected so as to take over the functions of the fallow and permit a form of crop rotation which increased rice yields, and which could also be produced with the rice growing equipment which was already available. The development towards crop rotation took the course shown in Table 22 especially on large farms.

In Stage 1 (I–II) of Table 22 the fallow lands were extended because the damage caused by monoculture was becoming increasingly obvious (single-crop farms). In Stage 2 (III) the land released by the reduction of the rice-growing areas was at first green manured (oats and lespedeza). However, the lespedeza was then extended to the fallow land until the government's assistance fund, which is limited by area, was exhausted (single-crop farms). In Stage 3 (IV) the land conservation crops—oats and lespedeza hay—were sold as cash crops (specialized farms). In Stage 4 (V), as

Table 22　Evolution of crop rotation on Arkansas rice farms.

| I | II | III | IV | V | VI |
|---|---|---|---|---|---|
| 1: Rice | 1: Rice | 1: Rice | 1: Rice | 1–2: Rice | 1: Rice |
| 2: Rice | 2: Rice | 2: Rice | 2: Rice | 3: Soya | 2: Soya |
| 3: Rice | 3: Fallow | 3: Green manure | 3: Oats | 4: Oats | 3: Oats |
| 4: Fallow | | (Oats and | 4: *Lespedeza* | 5: *Lespedeza* | |
| | | *lespedeza*) | | | |

| | | | | | |
|---|---|---|---|---|---|
| Rice as % of cultivated area | | | | | |
| 75 | 67 | 67 | 50 | 40 | 33 |

| | | | | | |
|---|---|---|---|---|---|
| Other crops as % of cultivated area | | | | | |
| – | – | 33 | 50 | 60 | 67 |

| | | | | | |
|---|---|---|---|---|---|
| Fallow as % of cultivated area | | | | | |
| 25 | 33 | – | – | – | – |

| | | | | | |
|---|---|---|---|---|---|
| No. of species | | | | | |
| 1 | 1 | 3 | 3 | 4 | 3 |

| | | | | | |
|---|---|---|---|---|---|
| No. of marketed products | | | | | |
| 1 | 1 | 1 | 3 | 3 | 3 |

Mono-product unit ————————————————————————→ Specialized unit

soya prices rose, farmers started to grow soya which requires the same equipment as rice. Oats and lespedeza hay were now used to fatten cattle since the market for them had proved too limited. Oats and lespedeza thus became green manure, marketable products, and fodder crops in succession (specialized farms). In Stage 5 (VI) lespedeza hay was finally discontinued because this fodder crop required additional machinery and its functions in the crop rotation were taken over by soya beans. Consequently, cattle farming was also discontinued and the whole farm specialized in the three marketed products: rice, soya and oats (specialized farms).

On smaller rice growing family farms in Arkansas development took a rather different course. Like the large rice farms they aimed at specialization but their point of departure in 1930 was not from single cropping but from mixed farming. Family farms were still diversified in 1930 because smaller farm units always tend to spread the economic risk and before mechanization, rice was very labour-intensive and could occupy only a small part of the farmland. Farms which were too small for share-cropping needed other activities to employ labour for the rest of the year. In addition smaller farms at that time still needed a considerable area of land for producing their own requirements in food and cattle fodder.

Mechanization eventually led to specialization in this case as well, because it needed large areas to maximize capital productivity. Mineral fertilizers and pest and weed control also relaxed the strict need for a crop rotation and mechanization levelled off peak demands for labour. Rice prices rose sufficiently to make it a more competitive crop as did technical progress in levelling, irrigation, drainage, storage, and drying. Finally farms became larger. In 1920 rice was usually grown once every seven years, but in 1960 this had increased to three times every five years or even twice every three years.

*Summary*

For approximately half of the world's population especially in the densely populated regions of South East Asia with their humid tropical climate, rice is sometimes the only basic food (Franke et al 1967 p.175). Rice is predominantly a subsistence crop, since only an average of about 15 to 25% of the rice harvest of tropical Asia leaves the farm and only about 3% of total world production is put on the market (Uexküll 1968 p.255). Consequently good rice harvests cause a strong pressure on prices. Uexküll illustrates this by pointing out that in Thailand the 1959 rice harvest was 38.5% higher than that of 1958, but its value at wholesale prices rose only 3%. In Indonesia, new, high-yielding varieties of rice which need large quantities of fertilizer have created local surpluses and led to reduction of up to 50% in producer prices.

Rice is grown under a large variety of economic and ecological conditions. It is found in every type of country, from agricultural to industrial, from the Equator to the plains of the river Po. The highest density of rice cultivation is, however, found in the overpopulated developing countries of the humid tropics. In Indonesia rice occupies 50.4% of the total cultivated area and in Thailand 49.5%.

Rice is cultivated by comparatively extensive methods in sparsely populated agricultural countries, by highly labour-intensive methods in densely populated agricultural countries, and by capital-intensive methods in industrial countries. Economic development is normally characterized first by an increase of population and later by industrial growth, and these in turn increase prices of land and of labour but decrease the price of capital. This same pattern, extensive ———▷ labour-intensive ———▷ capital-intensive, is reflected in the changing techniques of rice production as a country develops so that priorities for the productivity of capital of land and of labour are successively satisfied.

Rice production methods cannot be fully subordinated to economic requirements, because they are primarily dependent on natural conditions. All methods can be broadly summed up as those with and those without irrigation. At a world level irrigated production largely predominates.

These methods of cultivation also broadly determine the general cropping system. Rice grown without irrigation fits much better into a crop rotation than irrigated rice. Rice is often the only profitable crop for wet fields and has to remain a monoculture so that the expensive irrigation facilities are economically utilized.

At more advanced stages of economic development this restriction becomes less significant as farms and fields are larger and mechanization makes it cheaper to build dams. At this stage irrigated rice can begin to be grown in a crop rotation. Rotation is now more necessary because of the economic need for higher yields which require rotation to control diseases, pests, and weeds.

If one remembers that in industrial countries yield-increasing inputs such as chemical fertilizers and plant protection products, are plentiful and cheap, it is easy to understand why rice yields in industrial countries are 5000–5500 kg/ha, although in sparsely populated developing countries, they range only from 1400 to 2600 kg/ha, despite the fact that rice is sometimes grown with double or triple cropping in the developing countries while in most cases industrial countries are limited to single cropping.

## Comparison of the Economics of Sugar Beet and Sugar Cane Production in Khuzistan

### Introduction

The Iranian sugar industry has advanced and expanded considerably since the 1960s. As Fig. 23 shows the population of Iran grew by 30% between 1963 and 1972. Because of improvements in living standards sugar consumption per head also grew by almost 50%. Both developments together led to an increase in the total demand for sugar of about 99%. The gross value of sugar production, however, grew to about 272% so that imports of sugar could be reduced despite the rising population and increased consumption.

Fig. 23
Iranian sugar consumption and production 1963-1972 (1963=100). (Source: FAO 1972-73, UN 1973)

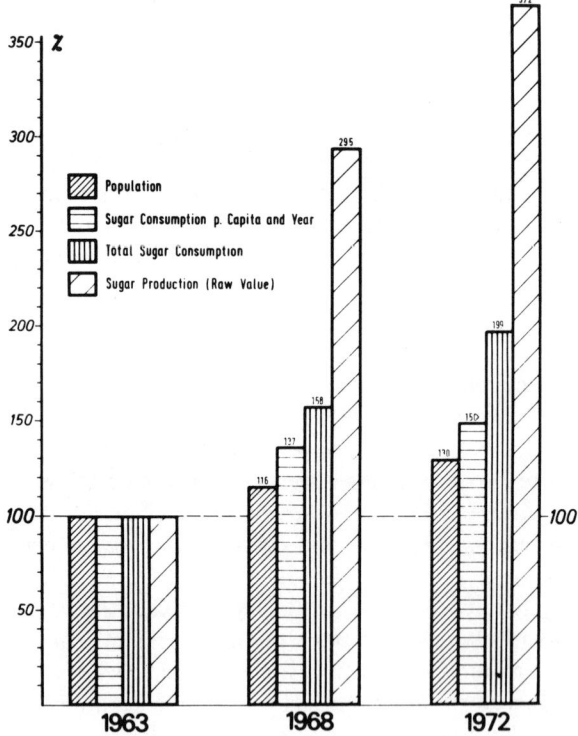

Despite this expansion in production, self sufficiency in sugar had only reached about 81% in 1974. In the effort to close the gap in self sufficiency attention was turned to the province of Khuzistan. Here notice had been attracted to two new methods of sugar production and here there were also considerable reserves of water which could be made available.

After the erection of the Dez dam, the largest storage dam in Asia, in the province of Khuzistan Iran planned to build further storage dams, in the first place to irrigate an additional 100 000 ha of desert steppe and largely to make it available for sugar production. The area involved was primarily the Ahvaz-Dezful region in the north of the province, latitude 31° to 32° N. with 220 to 250 mm of precipitation (from mid November to the end of March), at the most five humid months, a temperature of 15°–20°C in January and February, 20°–38°C from March to May, up to 50°C in summer and light frosts of 2°–3°C (occasionally up to 7°) in January and February. Land was in plentiful supply, water scarce. There was a danger of salting not from the water but from the land.

In this area it is both economically and ecologically possible to grow either winter sugar beet or sugar cane. Indeed in 1974 Khuzistan had both a beet sugar factory and cane sugar factory. I was asked to carry out an agricultural economic study as to which crop was the more suitable and showed the greater economic advantages. Khuzistan, like the Peshawar basin in Pakistan, Morocco, and the Andalusian Costa del Sol, is one of the few agricultural areas where sugar cane and sugar beet are competitive not merely from a general economic point of view but also at the farm management level. The question was therefore of scientific interest as well as of immediate economic importance. As these two irrigated crops are also beginning to become competitive in other parts of the sub-tropical arid belt in recent years, some of the results of my research are presented below.

## Economic comparison of winter grown sugar beet and sugar cane

Productivity of sugar beet and sugar cane in Khuzistan were first calculated and compared with results from the USA. As regards yields, experiments over a long number of years at the research station at Mardan in Pakistan had shown that in the Pershawar basin yields per ha from beet sugar were even greater than those of cane sugar (H. Blume 1967). The following extracts from the productivity comparison in Khuzistan are of interest:
1. The ratio of the annual sugar yield of beet to cane, 5.1:9.5 tonne/ha looks comparatively unfavourable for beet as average world yields in 1975/76 actually lay at 3.55:3.88 tonnes/ha raw sugar values.
2. As regards land productivity beet is 15% behind cane in Khuzistan, although in the USA the position is reversed and land productivity of beet is 20% higher than cane.
3. Water productivity of sugar beet is however 146% better than that of cane in Khuzistan.
4. Labour productivity of beet is also 34% higher in Khuzistan and up to 53% higher in the USA.

It is thus clear that sugar cane has relatively higher productivity in terms of land but that beet is superior in terms of water and labour productivity. In Khuzistan water is the most scarce factor of production and manpower is less plentiful than land. The productivity comparison for Khuzistan therefore comes down in favour of sugar beet. Under present economic conditions, therefore, the growing of sugar beet is to be preferred to growing cane. This applies at least to economies where gross productivity in monetary terms provides the criterion for decision.

Gross productivity is not however the only measuring rod of general economic benefit. The ideal is full cost accounting which in this case can also be carried out. The proportion of fixed costs to be taken into account is small because sugar cane plantations and sugar cane factories are almost always part state enterprises because of their monoproduction; practically no buildings are needed for growing beet; machine work is carried out by hired contractors and labour can be hired and fired as it is needed.

Only variable costs therefore have to be taken into account and these can be fairly easily compared. The internal farm costs are relatively easy to determine because free chippings are not returned, the leaf of the beet is mainly ploughed in and because of the low humus effect in the dry hot climate of Khuzistan the relative buying price of the fertilizer nutrient is easy to calculate. Even if the leaf is used as fodder it can be valued quite easily relative to the selling price of the widely used lucerne hay. The main difficulties in full cost accounting are in assigning farm management costs and in estimating the value of sugar beet as a preliminary crop in a rotation. This is clearly less than in a more temperate climate.

The model calculations I and II in Tables 23 and 24 and Fig. 24 show full cost accounts for growing winter sugar beet and sugar cane. These made it possible to derive the net productivities in Figure 25. These do not shift the gross productivity relationships between sugar beet and sugar cane in principle but only move it gradually so that our conclusions are not changed.

**Sugar beet and sugar cane from the point of view of general economic development**

Planning in the sugar industry must, however, go beyond current conditions and look far into the future. The equipment of an irrigation system, the creation of a particular infrastructure and agricultural structure, and the building of sugar factories with the right kind of transport arrangements all involve long term economic commitments. Before one can take an economic decision in favour of sugar beet on the basis of Figs. 24 and 25 one has to investigate whether economic forecasts show that sugar beet is likely to continue to have the advantage over cane in the foreseeable future.

An investigation by the US Department of Agriculture published in 1970 describes economic growth in Iran and other countries between 1950 and 1968. In Iran real income per head in this period showed the remarkably high annual growth rate of 3.5%. Inland demand for agricultural products grew by an annual rate of 4.7% as a combined effect of this 3.5% income growth, a 2.9% annual increase in population and an income elasticity of demand for agricultural products of 0.5. If this development continues in the foreseeable future it will affect the supply position of beet and cane sugar in the following three ways:

(1) The comparative labour productivity of growing beet will rise still further because of the rising wages to be expected.
(2) Sugar beet's smaller water requirements will be an advantage in the face of competition for scarce water resources from other economic sectors and the domestic population.
(3) Sugar beet rotations, with two crops a year, produce a higher total food supply than sugar cane. Byproducts like leaves, chips and molasses could produce either about 8000 kg of milk or about 730 kg of beef cattle (live weight) per ha of beet, even the leaves and tops alone could produce 5200 kg of milk or 395 kg of beef-cattle per ha.

In every area in the world where sugar beet is grown it has been shown that it

Table 23    Model calculation I. Costs, returns and profits on winter sugar beet in Rls/ha.

| | | Totals |
|---|---|---|
| 1.  Labour and machinery costs | | |
| 400 man hours at 13.5 Rls/mh | 5  363 | |
| tractors and machines | 7  359 | 12  722 |
| 2.  Inputs which increase yield | | |
| seed (10 kg at 110 Rls/kg) | 1  100 | |
| water (12  500 m$^3$ at 0.2 Rls/m$^3$) | 2  500 | |
| manure and green manure | 0 | |
| mineral fertilizers (240 kg N + 100 kg P$_2$O$_5$) | 6  986 | |
| plant protection materials | 1  280 | 11  866 |
| 3.  Other variable costs | | |
| rent | 1  550 | |
| interest on working capital (12%) | 500 | |
| insurance premium (legally compulsory) | 700 | 2  750 |
| 4.  Share of fixed cost, particularly management | | 2  000 |
| 5.  Gross costs of production | | 29  338 |
| 6.  Returns on byproducts (root: leaf = 1:0.5) | | |
| feed value of residues | 3  000 | |
| fertilizer value of the leaves ploughed in | | |
| (70 kg N + 21 kg P$_2$O$_5$) | 1  860 | |
| value as a preliminary crop | | |
| (0.9 + lucerne at 5  000 Rls) | 4  500 | 9  360[1] |
| 7.  Net production costs | | 19  978 |
| 8.  Transport costs to the factory (280 Rls/+ beet) | | 9  800 |
| 9.  Production and transport costs | | 29  778 |
| 10.  Returns on beet (35 t/ha at 1  800 Rls/ha) | | 63  000 |
| 11.  Profit | | 33  222 |
| profit as a percentage of returns = 52.7% | | |
| sugar content (in the sack) = 14.5% | | |
| sugar yield = 5.1 t/ha | | |
| total raw material costs up to processing = 5  800 Rls/t sugar | | |

[1] The leaves and tops from 1 ha of sugar beet have approximately the nutrient value of 3 tons of lucerne hay at 5  000 Rls/t = 15  000 Rls/ha allowing for the cost of dealing with the leaves.

promotes the intensification of agriculture and that it raises agricultural development to quite a new stage. Fig. 26 outlines the many ways in which sugar beet affects the farming system. Although the larger bulk of the plant material leaves the farm when the roots are sent to the factory much of it returns in the form of fertilizers (defecation scum) and feeding stuffs (molasses and dried pulp). These fodder byproducts combined with the leaves and tops of the plant provide a basis for a livestock enterprise. This in turn provides the manure to improve the fertility of the soil and the level of yields. Yields are also improved because of sugar beet's good effect on the next crop in the rotation or because the leaves may be ploughed directly back into the soil.

Table 24   Model calculation II. Cost, returns and profits on sugar cane in Khuzistan in Rls/ha.

| | | Totals |
|---|---:|---:|
| 1. Labour and machinery costs (including transport to the factory) | | |
| 630 man hours at 28 Rls/mh | 17 640 | |
| tractors and machines | 10 800 | 28 440 |
| 2. Inputs which increase yield | | |
| seed (new varieties) | 7 | |
| water (3600 m³ at 0.2 Rls/m³) | 7 200 | |
| manure and green manure | 0 | |
| mineral fertilizers (170 kg N + 116 kg P₂O₅) | 3 022 | |
| plant protection materials (esp. herbicides) | 1 280 | 11 509 |
| 3. Other variable costs (some items estimated) | | |
| rent | 7 000 | |
| interest on working capital (12%) | 600 | |
| reserve against risk (3% turnover) | 2 321 | |
| buildings | 1 381 | |
| management | 2 500 | 13 802 |
| 4. Share of overhead costs | | 1 800 |
| 5. Production and transport costs | | 55 551 |
| 6. Gross returns (100 t at 846 Rls/t) | | 84 635 |
| 7. Profit (when factory price of sugar is 22.5 Rls/kg) | | 29 084 |
| profit as a percentage of gross returns = 34.4% | | |
| sugar content (in the sack) = 9.5% | | |
| yield of sugar = 9.5 t/ha | | |
| total raw material cost up to processing = 5847 Rls/t sugar | | |

(Table values use LaTeX for formulas: water (3600 $m^3$ at 0.2 Rls/$m^3$); mineral fertilizers (170 kg N + 116 kg $P_2O_5$).)

Method: as returns and profits are only available for the combined plantation and manufacturing enterprise only one third of profits have been allocated to the plantation and two thirds to the factory in consideration of the low efficiency of agriculture as compared to industry. The national price of sugar cane is thus based on the sum of production and transport costs and of profit.

Sugar beet is not only an intensive crop itself but it is a crop which promotes intensification of the whole farming system. Sugar beet was therefore especially suitable to help achieve the 5.8% annual growth in agricultural production which was one of the most important targets of Iran's five year development plan from March 1973 to March 1979.

All efforts at economic development aim to increase real income per head, indeed this acts as a measure of all economic growth. As income per head rises the labour factor automatically becomes more expensive and scarcer. In agriculture this leads to competitive shifts in the course of economic development to types of farm enterprise which have higher labour productivity and which are best able to withstand the pressure of rising wages. As Fig. 25 shows, sugar beet has a 10% higher net labour productivity than cane and must therefore become increasingly superior to cane as economic development proceeds.

Fig. 27 shows how the costs of the raw materials for producing one ton of white sugar rise with rising wages and how much more sugar cane is affected by this than beet. Sugar cane is more sensitive to wage changes than beet, because of its higher input in manhours per ha. and because of its higher wage rates per manhour. Sugar beet production is comparatively less affected by wage rises.

Fig. 24 and 25
Sugar beet and cane, Khuzistan, 1972

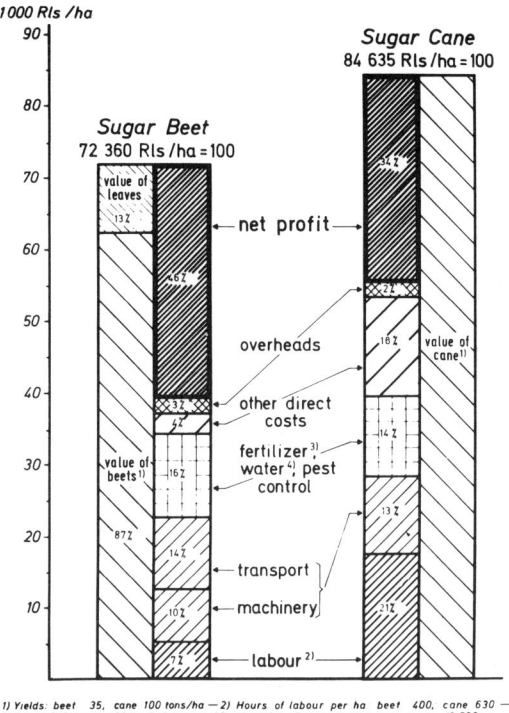

Fig. 24
Comparison of costs and profits

1) Yields: beet  35, cane 100 tons/ha — 2) Hours of labour per ha  beet  400, cane 630 —
3) Beet  240 kg N + 100 kg P₂O₅, cane 170 kg N + 116 kg P₂O₅/ha — 4) Water  beet  12 500,
cane  36 000 cbm/ha

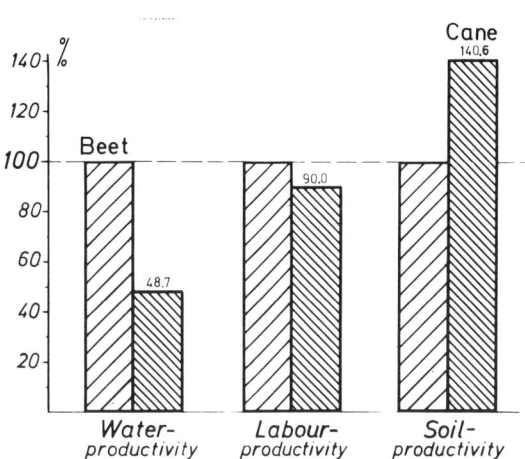

Basis of Calculation:

| Crop | Water used cbm/ha | Labour used h/ha | Yields tons/ha | Prices Rls/ton | Leaves Rls/ha |
|------|-------------------|------------------|----------------|----------------|----------------|
| Cane | 36·000 | 630 | 100 | 846 | — |
| Beet | 12 500 | 400 | 35 | 1800 | 9 360 |

Fig. 25
Comparison of productivity

Fig. 26
Sugar beet within the farming system. (Source: Hagelberg 1971, p.174)

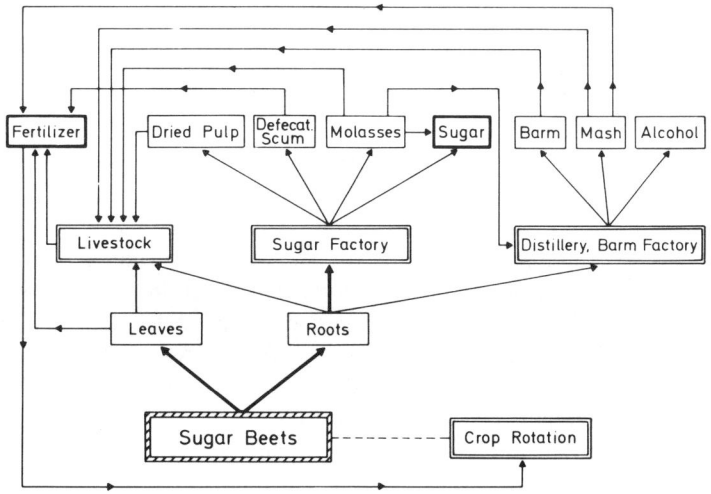

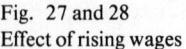

Fig. 27 and 28
Effect of rising wages

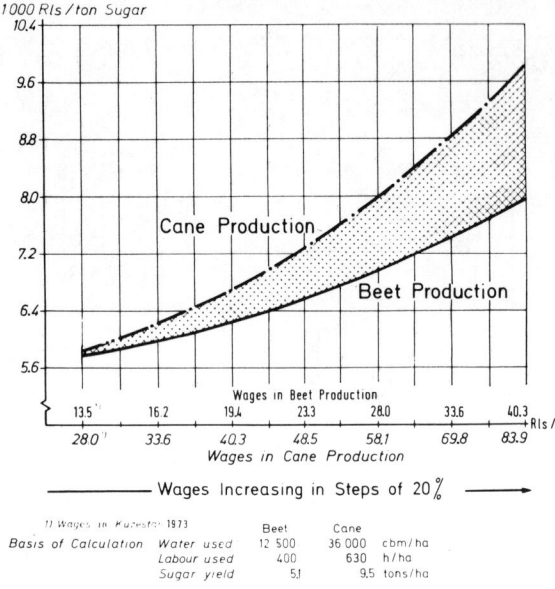

Fig. 27
Costs of cane or beet delivered to the factory per ton of sugar

Fig. 28
Profitability per ha of cane production as a percentage of
profitability of beet production

Fig. 28 shows a model calculation of how profits from the sugar beet and sugar cane are affected by rising wage levels. The byproducts are taken into account. With present wage levels the profits from sugar cane are only 87.7% of those from sugar beet. As wages rise the clear superiority of sugar beet increases. Thus if hourly wages were double present levels sugar cane profits would fall to only 41.0% of those of sugar beet.

Sugar beet has still further advantages during a period of economic growth. As demand and prices of milk and meat rise with increasing mass incomes the value of byproducts which can be used as fodder in the livestock sector also increases. But as beet produces much more of these fodder byproducts than cane its profit advantage over cane is even greater than that shown in Fig. 28. In Khuzistan the leaf alone at present accounts for 15–24% of the value of the beet (estimated in terms of costs of substituting lucerne).

Garrot (1971) calculated that between 1947/49 and 1967/69 the labour costs of producing a ton of cane sugar rose by 47% in Puerto Rico and by 34% in Hawaii. During the same period the labour costs of a ton of beet sugar produced on the American mainland fell by 4%. Sturrock (1969 1972) compared the development of total production costs of cane sugar in Jamaica with those of beet sugar in England over a 17 year period. He came to the conclusion that costs of growing, harvesting and transport to the factory for producing one ton of refined sugar had risen 86% for cane sugar between 1954 and 1970, while those for beet sugar had risen only 4%. Nothing can show more clearly how greatly the competitive advantage shifts towards sugar beet and away from sugar cane in the course of economic development.

**Combined factories minimize costs of sugar in Khuzistan**

Sugar beet provides the cheapest raw material for sugar production and in particular it provides higher profits in absolute and relative terms to its grower than sugar cane.

If one follows sugar production further to its end product, however, the heavier processing costs of sugar beet become apparent. This is due to the very short season of 60 to 70 days (see Fig. 29 below). The season is closely limited by climatic factors to between the end of the rainy period (clay soil) and the beginning of the hot period with temperatures up to 50°C which damage the quality of the beet. Processing must therefore be carried out in the period between the middle of April and the end of June. Beet sugar thus has low growing costs but high processing costs. Conversely cane sugar has higher growing costs but lower processing costs.

In principle processing costs for beet sugar could be reduced by lengthening the season. In theory this would be possible in three ways. (1) Summer beet could be brought in so that both a summer crop and a winter crop could be processed. This solution is not possible in Khuzistan because of its distance from the nearest area producing summer beet and the high transport costs that would be involved. (2) A storage system could be used for the beet that kept losses low. In the summer temperature of 50° in Khuzistan this is hardly possible. (3) Semi-processed sugar beet products could be stored throughout the harvest period. In the USSR processes such as steam extraction with the addition of lime and storage of the semi-finished product in briquette form or the drying of sugar beet chips were both unsuccessful. The degree of success of the experiment in silo storage of thick syrup which has been going on in America for the last 10 years has yet to be evaluated. The results from the USA as well as those of other experiments in Czechoslovakia are of little value for our purpose as the syrup would have to be stored under quite different climatic conditions in Khuzistan. Cold storage of syrup in the Khuzistan summer has yet to be tried and would certainly be very difficult. In addition the thick syrup processing system involves much more complicated technology than conventional methods. Whenever capacity was extended one would have to take account of costs of equipment for producing the syrup, capital and maintenance costs of storage silos, storage losses and interest charges on the syrup.

The most promising method seems likely to be the combined cane and beet sugar factory as this considerably lengthens the processing season and therefore brings down costs considerably.

The question, therefore, arises whether, in one of the few world agricultural regions where it is possible to grow both sugar bearing crops, combined processing can obtain the benefits of each crop while minimizing their disadvantages.

The conceptual model is a factory which although it would incorporate separate processing facilities for beet and cane would be able to make joint use of all other investment that was not dependent on the raw materials. Such a factory could process sugar cane for the six months from the end of September to the beginning of March and, after a one month break for maintenance, could process sugar beet from the end of April to the end of June. Sugar cane, which is difficult to transport should be grown in the area closest to the factory while sugar beet would be grown in the outer ring.

Fig. 29

Relation of sugar beet processing costs to length of season.
(Source: Strube & Brandes 1972)

Fig. 30

Sugar processing costs in Khuzistan in different types of
factories, in 1973 in Rls 1 000

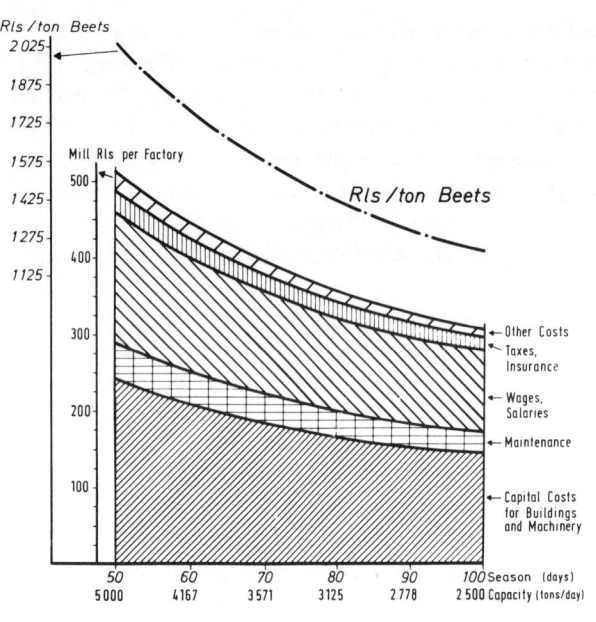

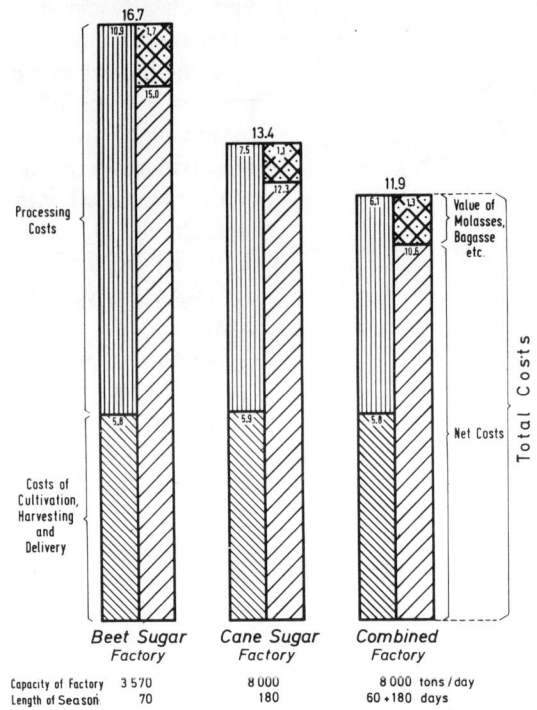

Fig. 30 shows that because of the extremely long processing season sugar processing costs of such a combined factory would be even less than those of processing sugar cane alone. Such a system would dispose of the final objection to expanding sugar beet production in Khuzistan, the high cost of processing.

The solution to Khuzistan's sugar production problem is therefore separate cultivation of cane and beet and combined processing. Even with separate processing winter grown sugar beet is one of the most paying farm enterprises for the Khuzistan farmer. It ranks next to the production of lucerne hay for sale, which at 13 cuts per year each yielding 18 tonnes worth R5000 produces a gross return of R90 000 per year. With combined processing, however, sugar beet also comes increasingly in the economic interest of the sugar industry and of the national economy in general.

Thus as further sugar processing plants are built in Khuzistan they should wherever possible have equipment provided for handling both crops. This will make sugar production as cheap as is possible. Combined processing should of course only be undertaken when existing processing plants are working to full capacity and the building of new factories is being considered. Thus when the Reza-Shah-Kabir storage dam makes it possible to grow sugar beet in a new irrigated area the possibility will only arise when the winter beet sugar factory in Ahvaz 60 km away can only deal with this production by extending its processing capacity. One of the oldest principles of economic behaviour is that one must first make the best possible use of existing resources before seeking to invest additional capital.

### Conclusions: how to expand sugar production in Khuzistan

Over the last 20 years Iran has been going through a period of uninterrupted economic development. The most important result of this development has been a rapid rise in real income per head which has led to rising labour costs for sugar production. The net productivity of labour for growing, harvesting and delivery to the factory is in the ratio of 100:90 in favour of sugar beet and against sugar cane. This alone makes sugar beet superior as a crop to sugar cane and this superiority will increase still further with further economic growth.

There are other advantages apart from labour productivity which winter sugar beet shows over sugar cane. With appropriate farming methods sugar beet can produce 5.8–7.2 t/ha of sugar while cane can produce 9–10 t/ha. However, the value of fodder by-products of beet are considerable. Leaves and tops produce about 2600 starch units per ha, sufficient to produce at least 5200 kg of milk or 395 kg of beef cattle. The value of these feedingstuffs to the livestock sector will continue to rise with economic growth, keeping step with the growing demand and rising prices for milk and meat. In addition it is possible to grow a second summer food or fodder crop in the same year after the winter beet has been harvested.

The 5.1 tonnes of sugar per ha produced from beet require 12 500 m$^3$ of irrigation water per ha while the 9.5 tonnes from sugar cane requires 36 000 m$^3$. This is ratio per ton of sugar in favour of beet of 154:100. This fact also means that the competitive position of beet will improve still further in the course of development as competing demands for water for domestic supplies, agriculture and industry reduce water reserves and make water more expensive.

Under the yield, price and cost conditions of 1973/74 the profit on growing beet was 33 222 R/ha and that sugar cane was 29 084 R/ha. Gross returns on beet were 52.7% and on cane only 34.4%. These figures cover growing, harvesting and transport to the factory.

Combined processing of winter beet and sugar cane in Khuzistan after the pattern in Andalusia or Peshawar would extend the working season of the factories to about eight months and would thus considerably reduce the cost of processing the raw material into sugar. Any new factories in Khuzistan should therefore, where possible, be dual purpose factories. This would make it possible to combine the high profits of producing sugar beet with better profits from processing because of the longer season. This would produce sugar at the least possible overall cost for growing and processing. Such considerations are, however, only relevant when it is no longer economical to increase the capacity of existing factories.

As the capacity of sugar factories is expanded costs of transporting raw materials rise although processing costs fall. Taking full account of these conflicting factors and of the fact that all transport costs tend to fall in the course of economic growth, the setting up of large factories with a daily capacity of 3 000 t of beet is recommended. The relatively short beet processing season is taken into account in this recommendation. The relatively long interval of four to five years between sugar beet crops in a rotation is considered in estimating the collection area. Total procurement cost per ton of capacity sinks by 35–40% as daily capacity is increased from 1000 to 3000 tonnes of beet. There is scarcely any further reduction as capacity is increased beyond 3000 tonnes (Strube & Brandes 1972 p.396).

This problem of the competition between sugar cane and winter sugar beet is not special to Khuzistan. It is more than likely that advances in winter beet production methods will gradually drive sugar cane out of parts of the subtropical dry belt of the old world. The more rapid the general economic development of a country is the more rapid this process is likely to be, because the advantages of sugar beet over sugar cane increase with increasing economic development. Thus if wages per hour in Khuzistan were doubled the profit on sugar cane would cease to be 87.7% of that on beet and would sink to 41.0%. Developments in winter sugar beet production have made the sub-tropics a legitimate production zone for sugar beet and will to some extent drive sugar cane production into the humid tropics nearer the equator. The introduction of sugar cane into the sub-tropical dry belt and into Haft Tappets was a correct decision in its time because there was then no alternative way of producing sugar under these natural conditions. At the present stage of economic and technical development, however, sugar beet should have priority.

While both sugar cane and sugar beet are very profitable crops in Khuzistan, sugar beet is by far the more profitable. This sets future priorities because the better is the enemy of the good. Sugar cane production should, however, be planned in relation to sugar beet production in such a way as to gain the advantage of combined very low cost processing. The area relationship between winter beet and sugar cane for such combined processing lies at about 1:1.8 where sugar beet yields are around 40 t/ha and when the season lasts about 60 days or at about 1:1.5 when the season extends to 75 days.

The highest profits as far as raw material production goes lie with winter sugar beet, the most profitable sector of the sugar processing industry is the cane sugar sector. Beet provides the cheapest raw material, cane sugar is cheapest to process. The lowest overall cost of producing sugar is achieved when both crops are grown separately and processed in a combined factory.

There are however a number of important qualifications to this positive evaluation of combined beet and cane processing. As there are considerable technical problems in the production and storage of thick syrup, so there are a number of very important organisational problems to be solved in a combined beet and cane factory. In view of the fact that the Iranian government has long reserved cane sugar production and processing in its own hands the necessary investment for a combined processing factory and related cane sugar plantations is unlikely to be forthcoming from the private sector. On the other hand it is clearly in the national economic interest that the sugar beet sector should remain in private ownership. A combined factory would therefore involve a joint capital investment from the government and private sector which could raise considerable problems in distributing costs and profits. If these problems cannot be solved none of the methods we have suggested for lengthening the season for winter beet would be economically feasible under conditions in Khuzistan.

If this is the case, after weighing up all the pros and cons the following solution is proposed as the most rational to clear the way for the beet sugar production which is absolutely indispensable for the future development of Khuzistan and of the Iranian sugar industry. There should be a premium on the price of sugar delivered from winter sugar beet factories at a level sufficient to balance the seasonally conditioned cost disadvantage of winter beet factories as against those of summer beet factories. One is reluctant to make political recommendations to countries outside ones own. Nevertheless such compensation for cost disadvantages is a tested, legitimate and much used instrument of regional economic policy in nearly all countries. It is particularly suitable for encouraging development in less favoured economic regions as Khuzistan is in this case. Apart from direct price premiums the cost disadvantage of winter beet factories could also be compensated for by credit policy measures, for instance by guaranteeing an adequate amount of long term low interest credit. It would also be good for the economic policy of Iran to encourage the setting up of more agro-industrial corporations by preparing and leasing irrigated land and by low interest long term credits. The agro-industrial corporations are the route makers in winter beet production and this along with petroleum production holds the key to development in Khuzistan.

The enormous economic opportunities which winter sugar beet offers to Khuzistan must not be thrown away by immobilizing private initiative. The technological problems of storing the thick syrup can perhaps be solved by small scale research. The organizational problems of dual purpose factories must also be dealt with. At the very least there is no need to shrink from an economic policy which compensates for the cost disadvantages at the processing stage.

## More Efficient Use of Fresh Water Resources

### Water as the key to food production

In the first half of this century development in world food production was characterized by an increase in the use of fertilizers. In the second half of the century the main emphasis is on development of irrigated farming. The development possibilities would be unlimited if man were to achieve complete control of the water factor. Already the technical possibilities are very great but the ratio of costs to benefits has so far greatly restricted the large-scale use of many processes.

Ruthenberg (1967 p.153 et seq.) describes the effects of irrigation as follows:

1. The increase in gross returns per ha through increased yields per crop, and several crops per year especially in the subtropics and outer tropics, and a switch to higher-yielding types of crop, with intensified use of labour and fertilizers.
2. Achievement of continuous land use in tropical areas without intermediate periods of woodland bush, grass or fallow.
3. Reduction of fluctuations in harvests and consequently more regular food supplies for men and animals.
4. Increased flexibility in the programme of production and degree of intensity.
5. An increase in the viability of food production so that the minimum farm size can decrease.
6. An extension of farming to arid areas like wild savanna, dry steppe and semi-desert which could not otherwise be cultivated.

Water is however badly distributed. Some 80% of the African farming area suffers either from excessive humidity near the equator, or from periodic or permanent water shortages in the outer tropical zones. It is not always possible to balance the water surpluses against the shortages except where there are large rivers available to carry water from the wet areas to the dry ones.

Irrigation appears particularly promising in the tropical and subtropical areas away from the equator where precipitation is minimal or very sporadic. Here surface or sprinkler irrigation combined with the favourable heat conditions would permit several harvests per year, or even continuing crop production throughout

E

the year. Unfortunately water is particularly short in these dry climates where the rainfall is very low, there are virtually no rivers or lakes, and the groundwater level is also low. The water supply is therefore most inadequate in the very places where irrigation would produce the most valuable results. The areas most suitable for irrigation are the arid zones near mountains like the Andes, Atlas, Caucasus, Elburz, Zagros, Drakensberg Mountains or Ethiopian Highlands or zones linked to such mountains or to wet tropical areas by large rivers like the Nile, the Orange River, the Cunene, Tigris, Euphrates, Indus, or the Colorado River, which flow throughout the year.

As the gap between the demand for water and its supply generally increases with the dryness of the climate, so it also increases in the course of national economic development. This is not simply or even primarily because of increased irrigation and increasing demand for water for agriculture, but much more because of the rapidly increasing demand for water from other sectors of industry and for domestic usage.

Press (1959) points out that only about 40 litres of water per caput per day are required by a population supplied from wells. Once central waterworks are installed, however, summer consumption rises to 200–350 litres or more per caput per day. Industry requires up to 20 $m^3$ of water to produce a tonne of steel, 300 $m^3$ to produce a tonne of wood pulp, and as much as 800 $m^3$ to produce one tonne of rayon. Production of one tonne of vegetable matter requires about 500 $m^3$. In West Germany, agriculture, horticulture and forestry at present consume about 17 times as much water as industry and domestic users.

There are, however, various ways in which the management of water resources can be improved despite the steadily decreasing reserves.

### Replacement of surface irrigation by sprinkler irrigation

Sprinkler and trickle irrigation make much more productive use of water than surface irrigation. Thus cotton grown in Morocco with sprinkler irrigation requires only one third of the water needed when it is grown with surface irrigation (J. van Reveren 1959, page 14). Bandini (1959) states that in Italy sprinkler irrigation requires only half the volume of irrigation water otherwise required and in exceptional cases only one fifth. Sprinkler irrigation can, however, only be applied to projects at an advanced-stage of industrial development since it requires higher capital investment and lower labour expenditure than traditional surface irrigation. A second important difference in the cost structure is that it is the variable costs which are increased for sprinkler systems, while it is the fixed costs which are high for surface systems. This factor tends to restrict surface irrigation to the worlds arid zone where large volumes of water have to be supplied systematically. In contrast, sprinkler irrigation is more appropriate to climates where irrigation is used only to offset risk in occasional dry years. The water saving effect of sprinkler irrigation is thus only achieved in practice under particular economic and ecological conditions.

### Limiting non-productive evaporation of water

Water utilization can be improved if the soil surface is covered with a mulch or plastic foil so as to conserve the natural or artificial soil moisture. Special crops can be grown under optimum production conditions by laying a layer of foil or plastic or by spraying foam carpets. These methods conserve soil humidity, absorb heat and suppress weeds. Yields in glasshouses and temporary buildings made from foil and other plastics can be further increased through $CO_2$ enrichment.

In Hawaii the annual rainfall is only 650 mm, whereas the evaporation potential of a pineapple crop is 2030 mm. Hawaii nevertheless supplies some 80% of total world consumption of tinned pineapple, simply through making up the immense water deficit by means of large-area sprinklers. Black plastic foil, laid by machine, also saves expensive water and concentrates additional heat in the soil.

The situation is different when the soil is provided with a living cover of vegetation. Fruit trees grown on grassland (green cultivation) require more water than those on bare earth (black cultivation), since the evaporation of water from the living vegetation is greater than from the bare ground. If this were not so, there would be no point in operating the dry farming system. General economic development results in increasing shortages and higher costs first of labour and then of irrigation water but makes capital goods become cheaper. Thus, in climates of moderate humidity, the sequence of development in orchards and gardens must be: (1) Labour- and water-intensive black cultivation with cultivation of the soil; (2) green cultivation which saves labour at expense of increased water consumption; and (3) black cultivation using herbicides and sprinkle irrigation. This achieves very considerable savings in water and in labour at the expense of high expenditure on materials.

In Australia and Israel experiments are being carried out to reduce evaporation of water from the surfaces of lakes by spraying the surface with a film of organic chemicals. This technique is also being considered over the sea combined with artificially produced haze layers.

## Double utilization of fresh water with aquatic cultures

Damming as used for growing wet rice takes extremely large volumes of water. This water can sometimes be used for more than one purpose. Thus in Arkansas the rice fields are reflooded after harvesting, so as to attract the wild duck which fly past in the autumn on their way from Canada to the Gulf Coast. The capture of thousands of wild duck every year as an ancillary product to the rice, increases the productivity of the irrigation water. Wet rice production is also combined in various ways with fish farming, as in Java or Madagascar. The fish may either be reared in the growing rice field (yield up to 50 kg of fish per ha) or the rice field may be stocked with fish after the rice has been harvested (yield up to 300 kg of fish per ha) (Ruthenberg 1967 page 153).

Double use of fresh water is simply a combination of agriculture and aquatic culture. As agriculture is the generic term for any method of food production on land, aquatic culture (nowadays often called aquaculture) covers all the techniques for cultivating economically valuable water organisms.

The productivity of water used in fish ponds may also be greatly increased in the future. In the U.S.A. in 1970 some 35 000 tonnes of catfish were caught from 2300 ha of flooded fields (yield 1.52 t/ha). The food return is twice that of poultry. Only 1.5 kg of special pellets are required per kg of catfish.

In Germany aquaculture as a technique of intensive fish production is particularly associated with a method developed at Ahrensburg near Hamburg. Carp are reared in continuous-flow aquaria in glasshouses operating on a warm water circulation system which gives year round accelerated growth. The warm-water carp increases its weight continuously while the pond carp grows only seasonally. Total independence of climate makes it possible for the fish in warm water to be bred, reared, and fattened throughout the year. Of two carp from a given pond population, one reared in the pond itself, weighed 40 g after one year, while another reared in warm water weighed 1750 g. This Ahrensburg system has intensified, even revolutionized the old fashioned conventional technique of fish farming.

The importance of this development work is highlighted because freshwater fish could soon be one of the main sources of protein. Between 1948 and 1975 world fish catches rose from 19.6 to 69.7 million tonnes (355%). It is questionable whether the forecast of a further increase to about 100 million tonnes by the year 2000 will be achieved. The indications are that rapid sophistication of the fishing fleet through fitting-out with radar and loran equipment, echo probes, fish locators, etc is no longer affording the sea fish the time and space it requires for regeneration.

The latest trends in sea fishing such as the herring shortage in the North Sea, the Icelandic fishery war, the disappearance of shoals of fish off Peru with consequent 20% or so decrease in world fishmeal supply, are typical results of modern methods and of the increasing contamination of water. Contamination of seas, rivers and ponds by introduction of toxic substances can have three different consequences for fish that are economically useful. (1) There can be sudden mass death through acute poisoning such as occurred in the Rhine in 1969. (2) The toxic substances can act on the fish eggs and larval stages leading to a sharp drop in the stock of sea fish. Thus Portugal's exports of canned sardines dropped almost 50% between 1968 and 1970 because of reduced catches. (3) Non-toxic storage of toxic substances in economically useful fish has caused symptoms of illness in humans eating fish in Scandinavia and has even led to deaths in Japan.

Eels as well as carp and trout, can now be bred by aquaculture. In Ahrensburg, 3500 kg of fish are kept in 350 million $cm^3$ of conditioned water which is circulated and clarified biologically. A concentration of only 100 litres of water per kg of fish is a spatial concentration similar to that in hatcheries. It offers similar advantages, increase in weight throughout the year, maturity of the female after only 15 months instead of 3 to 4 years under natural conditions, egg production throughout the year, and high feed conversion.

Mussel farming in the "Rias" of Galicia in north-west Spain also gives a high utilization of water. Suspended cultures of mussels are grown on ropes suspended from rafts. The cultivation of the common mussel in the Bay of Vigo is even more economical. Finally mention should be made of the algae farms, which have already achieved spectacular success using fresh water tanks and added nitrogen. Since the green algae provide primary production through photosynthesis, they only require a solution of mineral-substance in addition to light and carbon dioxide. Algae cultures, especially of chlorella, have been developed in Japan, the USA, France and the USSR. Fundamental work has been carried out in Germany on industrial mass production of the single-cell green alga Scenedesmus obliquus. This contains not less than 50–59% raw protein in the dry substance compared with 34-50% in soya. This alga is used for rat, pig and fish food. It is also suitable as a human foodstuff.

Tropical forests also produce considerable useful wastes for breeding of single-cell protein (bacteria, yeast, single-cell algae, etc.). Conditions for the production of single-cell protein are therefore favourable in the area where the effects of protein deficiency on human nutrition are the greatest.

# Bibliography

Andreae, B. (1961) [Cultivation with irrigation. Views on location of spray irrigation and trickle irrigation.] Landbau mit künstlicher Bewässerung. Betrachtungen zur Standortsorientierung von Beregnung und Berieselung. In: Beiträge zur landwirtschaftlichen Betriebslehre. Festgabe zum 65. Geburtstag von Professor Dr. Dr. H.c. Georg Blohm. Stuttgart, German Federal Republic; Verlag Eugen Ulmer 10–36.

Andreae, B. (1964) [Farm types in agriculture. Origin and change in land use, livestock husbandry and farming systems in Europe and abroad, and new methods of defining them. Methodological part of a text book on farm management.] Betriebsformen in der Landwirtschaft. Entstehung und Wandlung von Bodennutzungs-, Viehhaltungs- und Betriebssystemen in Europa und Übersee sowie neue Methoden ihrer Abgrenzung. Systematischer Teil einer Agrarbetriebslehre. Stuttgart, German Federal Republic Verlag Eugen Ulmer 426pp.

Andreae, B. (1965) [Land fertility in the tropics. Utilization and maintenance. Farm management considerations for work in developing countries.] Die Bodenfruchtbarkeit in den Tropen. Nutzbarmachung und Erhaltung. Betriebswirtschaftliche Überlegungen für die Arbeit in Entwicklungsländern. Hamburg, German Federal Republic; Paul Parey 124pp.

Andreae, B. (1975a) Increasing food production through more efficient use of water. *Plant Research and Development* 2, 122–133.

Andreae, B. (1975b) Types of irrigation farming. *Applied Sciences and Development* 6, 77–93.

Andreae, B. (1977) [Alternatives to sugar production in the Peshawar Basin of Pakistan.] Alternativen der Zuckerproduktion im Peshawar-Becken Pakistans. *Zeitschrift für die Zuckerindustrie* 27, 89–93.

Aurada, F. (1960) [Irrigation systems of the Indus Lowlands and problems in their development.] Bewässerungssysteme des Industieflandes und ihre Entwicklungsprobleme. In: Mitteilungen der Österreichischen Geographischen Gesellschaft. Vienna, Austria 326–339.

Bandini, M. (1959) [Agricultural economics.] Economia Agraria. Torino, Italy; Unione Tipografico-Editrice Torinese 756pp.

Barker, R.; Anden, T. (1975) Factors influencing the use of modern rice technology in the study area. In: Changes in rice farming in selected areas of Asia. Los Baños, Philippines; International Rice Research Institute.

Barnes, A.C. (1964) The sugar cane. London, UK; New York, USA; Hill and Interscience Publication 456pp.

Benneh, G. (1973) Small-scale farming systems in Ghana. *Africa* 43, 134–146.

Beveren, J. van (1959) [Results of irrigation in tropical and sub-tropical regions of Africa.] Erfolge der Beregnung in den tropischen und subtropischen Gebieten Afrikas. In: Beregnungstechnik von heute und morgen. German Federal Republic; Mannesmannregner-GmbH p.15 et seq.

Biehl, M. (1973) [Agriculture in China and India.] Die Landwirtschaft in China und Indien. Frankfurt/Main, German Federal Republic; Movitz Wiesterweg (1973) 119pp. Ed.4.

Blume, H. (1967) [Sugar cane and sugar beet in the dry sub-tropical belt of the Old World.] Zuckerrohr und Zuckerrübe im subtropischen Trockengürtel der Alten Welt. *Erdkunde* (1967) 21, 111–132.

Booker, L.J. (1952) A suggested method of comparing costs of surface versus sprinkler irrigation. (Unpublished Manuscript, University of California).

Bradfield, R. (1972) Maximizing food production through multiple cropping systems centered on rice. In: Rice, science and man. Los Baños, Philippines; International Rice Research Institute 143–163.

Breth, S.A. (1972) IRB and beyond. Los Baños, Philippines; International Rice Research Institute.

Caesar, K. (1960) [Basis and structure of Iraqi agriculture.] Grundlagen und Struktur der irakischen Landwirtschaft. *Berichte über Landwirtschaft* 38, 188–208.

Caesar, K. (1968) [Distribution of crops over the world's climatic zones.] Die Verbreitung von Kulturpflanzen in den Klimazonen der Erde. *Umschau in Wissenschaft und Technik* 13, 391–397.

Christiansen-Weniger, F. (1970) [Forms of arable farming in the Mediterranean area and the Near East.] Ackerbauformen im Mittelmeerraum und im Nahen Osten. Frankfurt/Main, German Federal Republic; DLG-Verlag 500pp.

Clark, C. (1960) The economics of irrigation in dry climates. Oxford, UK; Agricultural Economics Institute, Oxford University 31pp.

Decken, H. von der (1957) [Agriculture and the food sector in Egypt.] Landwirtschaft und Ernährungswirtschaft in Ägypten. *Berichte über Landwirtschaft* 35, 688–716.

Delavier, H.J. (1967) [Sugar cane—sugar beet—a comparison.] Zuckerrohr—Zuckerrüben—ein Vergleich. *Internationales Zuckerwirtschaftliches Jahr- und Adressbuch* 1967. 229–233.

Dietz, P.M. (1975) Project accounts on sugar-cane. Chanray-Tinachenos, Peru. (Manuscript).

Dozina, G.E.; Herdt, R.W. (1974) Upland rice farming in the Philippines. Los Baños, Philippines; International Rice Research Institute (Mimeograph).

FAO (1972; 1973; 1976; 1977) *Production Yearbook* 25; 26; 29; 30.

Feistritzer, W.P. (1969) Agricultural research and propagation of improved seeds. *Almanach 1969 der Deutschen Stiftung für Entwicklungsländer.*

Fels, E. (1954) [Man's economic activities shape the earth.] Der wirtschaftende Mensch als Gesalter der Erde. In: Lütgens, R. (*Ed*) Erde und Weltwirtschaft, Vol.V. Stuttgart, German Federal Republic; Frankh'sche Verlagshandlung 258pp.

Finck, A. (1970) [Food production potential in agriculture.] Möglichkeiten der Nahrungsproduktion im Landbau. *Ernährungs-Umschau* 2, 47–52.

Fischer, K.H. (1957) [Agriculture in Iraq.] Die Landwirtschaft im Irak. *Berichte über Landwirtschaft* 35, 456–471.

Franke, G. et al (1967) [Useful crops of the tropics and sub-tropics.] Nutzpflanzen der Tropen und Subtropen, Vols. I and II. Leipzig, German Democratic Republic; S. Hirzel Verlag 324pp.; 421pp.

Garrot, W.N. (1971) Labor productivity on sugarbeet and sugarcane farms in the United States 1946–69. *Sugar Reports* No.225, 20pp.

Gerling, W. (1954) [The plantation] Die Plantage. Ed.2, revised. Würzburg, German Federal Republic; Stahel'sche Universitätsbuchhandlung 47pp.

Gibb, Sir Alexander and Partners (1956) Water resources survey of the Nile Basin in Tanganyika; report. Dar es Salaam, Tanganyika and London, UK.

Gnielinski, S. von (1968) [Sugar cane growing in Liberia and its economic significance.] Zuckerrohranbau im Liberia und seine wirtschaftliche Bedeutung. *Zeitschrift für ausländische Landwirtschaft* 7, 276–291.

Gregor, H.F. (1976) Agricultural intensity in the Pacific Southwest. In: Proceedings, Association of American Geographers.

Hagelberg, G. (1969) [Structural changes in the world sugar industry since 1960.] Strukturwandlungen in der Weltzuckerwirtschaft seit 1960. *Jahrbuch für Wirtschaftsgeschichte* 4, 99–115.

Hanrath, J.J. (1964) [Economic problems of arid and semi-arid regions.] De economische problematiek der aride en semi-aride gebieden. *Tijdschrift van het Koninklijk Nederlandsch Aardrijkskundig Genootschap* 81, 172–181.

Harrison Church, R.J. (1961) Problems and development of the dry zone of West Africa (with discussion). *Geographical Journal* 127, 187–204.

Harwood, R.R.; Price, E.C. (1975) Multiple cropping in tropical Asia. Los Baños, Philippines; International Rice Research Institute. (Mimeograph).

Heady, E.O.; Jensen, H.R. (1954) Farm management economics. Englewood Cliffs, USA; Prentice-Hall Inc. 645pp.

Healey, D.T. (1964) Agricultural economics in some African countries. *International Journal of Agrarian Affairs* 4 (4), 250–286.

Heath, R.G. (1961) Reports to the Government of Ghana on crop production possibilities under conditions of irrigation in the Volta flood plain area. *Expanded Technical Assistance Programme, FAO Report* No.1404, 25pp.

Humbert, R. (1968) The growing of sugar cane. Amsterdam, Netherlands; London, UK; New York, USA; Elsevier Publishing Co. 780pp.

Institut für Ausländische Landwirtschaft der TU Berlin (1969) Basic agricultural data of developing countries. *Zeitschrift für Ausländische Landwirtschaft* 8 (1) 4pp.

IRRI (1975) Research highlights for 1974. Los Baños, Philippines; International Rice Research Institute.

Jamlekha, K.O. (1957) A study of the economy of a rice growing village in Central Thailand. Dissertation, Ann Arbor, Michigan 420pp.

Jensch, G. (1969) [Global climate.] Klima-Globus. Berlin; Riepert KG 50pp.

Jones, J.O. (1963) The economics of water supply and control. Conclusions. *International Journal of Agrarian Affairs* 4 (1), 50–58.

Jung, H.F.; Scheinpflug, H. [Rice cultivation and plant protection problems in Japan.] Reisanbau und seine Pflanzenschutzprobleme in Japan. *Pflanzenschutz-Nachrichten Bayer* 23 (4), 243–271.

Kemmler, G. (1968) [Fertilizing paddy-rice—experience from South and East Asia.] Die Düngung von Sumpfreis. Erfahrungen aus Süd- und Ostadien. *Tropenlandwirt* 69, 8–26.

Khan, A.R. (1972) The economy of Bangladesh. London, UK; Macmillan 196pp.

Könnecke, G. (1967) [Crop rotations.] Fruchtfolgen. Berlin; VEB Deutscher Landwirtschaftsverlag 335pp.

Krug, G. (1957) [The full rice bowl.] Die volle Schale Reis. *Westermanns Monatshefte* 5, 39.

Kung, P. (1962) Irrigation farming and multiple cropping in the Ganges-Kobadak project areas. *FAO Report* 1956.

Kürten, P. (1954) [Rice. Cultivation and fertilizer application outside East Asia.] Reis. Anbau und Düngung ausserhalb Ostasiens. *Schriftenreihe über Tropische und Subtropische Kulturpflanzen* 126pp.

Licht, F.O. (1967) [World sugar statistics 1966/67.] Weltzuckerstatistik 1966/67. *Internationales Zuckerwirtschaftliches Jahr- und Adressbuch* 350pp.

Müller, G. (1954) [Sugar cane, cultivation and fertilizer application.] Zuckerrohr. Anbau und Düngung. *Schriftenreihe über Tropische und Subtropische Kulturpflanzen* 109pp.

Narkswasdi, U. (1963) Farm management problems in Thailand. *World Crops* 15.

National Council of Applied Economic Research (1959) Criteria for fixation of water rates and selection of irrigation project. New Delhi, India.

Nitz, H.J. (1974) [Rice polders in southern Kerala (India).] Reislandpolder in Südkerala (Indien). *Heidelberger Geographische Arbeiten* 40, 443–454.

Nye, P.H.; Greenland, D.J. (1960) The soil under shifting cultivation. *Technical Communication, Commonwealth Bureau of Soils* No.51, 156pp.

Okigbo, B.N. (1974) The IITA farming system program. Ibadan, Nigeria; IITA (Mimeograph).

Olivier, H. (1961) Irrigation and climate. New aids to engineering planning and development of water resources. London, UK; Edward Arnold 250pp.

Piekenbrock, P. (1958) [Vegetation and crop cultivation in the tropics.] Vegetation und Pflanzenbau in den Tropen. *Schriftenreihe der Deutschen Afrika-Gesellschaft* No.7, 36pp.

Pilhofer, H. (1966) Studies in East Indian jute and paddy farms over time. *International Journal of Agrarian Affairs* 5 (1), 62–84.

Press, H. (1959) [Maintenance and reclamation of cultivated land.] Kulturlanderhaltung und Kulturlandgewinnung. Hamburg, German Federal Republic; Paul Parey 372pp.

Roemer, T. (1950) [Will the teaching of Robert Malthus (1798–1805) become reality in the second half of the 20th century after all?] Wird die Lehre von Robert Malthus (1798–1805) in der zweiten Hälfte des 20. Jahrhunderts doch noch Wirklichkeit? In: Gedenkschrift zur Doppel-Verleihung des Justus-von-Liebig Preises 1949 und 1950. Hamburg, German Federal Republic; Stiftung FVS zu Hamburg.

Ruthenberg, H. (1967) [Ways of organizing land use and livestock farming in the tropics and sub-tropics demonstrated by selected examples.] Organisationsformen der Bodennutzung und Viehhaltung in den Tropen und Subtropen, dargestellt an ausgewählten Beispielen. In: Von Blanckenburg, P.; Cremer, H.D. (*Eds*) Handbuch der Landwirtschaft und Ernährung in den Entwicklungländern, Vol.I. Stuttgart, German Federal Republic; Eugen Ulmer 122–128.

Ruthenberg, H. (1976) Farming systems in the tropics. 2nd edn. Oxford, UK; Clarendon Press 366pp.

Sapper, K. (1932) [Distribution of irrigation.] Die Verbreitung der künstlichen Feldbewässerung. *Petermanns Mitteilungen* 78, 225–231, 295–301.

Scholz, U. (1977) [Minang Kabau. Agricultural structure in West Sumatra and development potential.] Minangkabau. Die Agrarstruktur in West-Sumatra und Möglichkeiten ihrer Entwicklung. *Giessener Geographische Schriften* No.41, 217pp.

Schulenburg, Graf von der, W. (1960) [Business management and structural problems of the sugar industry in the German Federal Republic.] Betriebswirtschaftliche und strukturelle Probleme der Zuckerindustrie in der Bundesrepublik Deutschland. *Agrarwirtschaft* Spec. Issue 9 143pp.

Strube, C.; Brandes, W. (1972) [Model analysis to determine the long-term optimum length of seasons for beet sugar factories.] Modellanalyse zur Bestimmung der langfristig optimalen Kampagnedauer bei Rübenzuckerfabriken. *Agrarwirtschaft* 21, 396pp.

Sturrock, F. (1969) Sugar beet or sugar cane? *Cane Farmer* No.4, 104pp.

Sturrock, F.; Thompson, M. (1972) Sugar beet. A study of sugar production in the UK and the feasibility of expansion. *Economic Report Agricultural Enterprise Studies in England and Wales* No.7, 22pp.

Take, E.F.; Stepanow, A.S.; Ali, L.; Hallmanns, W. (1963) The world sugar economic structure and policies, Vol.II. London, UK; International Sugar Council.

Tsuzuki, T. (1963) [Crop rotation in Japanese arable farming.] Die Fruchtfolgen des Japanischen Ackerbaues. *Berichte über Landwirtschaft* 41, 837–846.

Uexküll, H.R. von (1969) [Rice in Asia—problems and possibilities of raising production.] Reis in Asien—Probleme und Möglichkeiten einer Produktionssteigerung. *Zeitschrift für Ausländische Landwirtschaft* 8, 248–259.

Uhlig, H. (Editor (1975) [Southeast Asia—the south Pacific area.] Südostasien—Austral-pazifischer Raum. Frankfurt/Main, German Federal Republic; Fischer Taschenbuch Verlag 491pp. Fischer Länder Kunde No.3.

UN, Statistical Office (1977) Statistical Yearbook 1976. New York, USA.

UNESCO (1960) Plant-water relationship in arid and semi-arid conditions. Reviews of research. *Arid Zone Research* 15, 225pp.

USA, Department of Agriculture (1970) Economic progress of agriculture in developing nations 1950–68. *Foreign Agricultural Economic Report, Economic Research Service* No.59, 180pp.

Vos, J.H. de (1938) [Changes in the location of sugar cane growing.] Standortswanderungen der Zuckerrohrkultur. Zürich, Switzerland.

Wahlen, F.T. (1956) [World agricultural problems as seen by the FAO.] Landwirtschaftliche Weltprobleme im Gesichtswinkel der FAO. *Berichte über Landwirtschaft* 34, 176–185.

Wang, Y.; Nagel, F.; Ruthenberg, H. (1969) [Land use and technical progress in Taiwan.] Bodennutzung und technischer Fortschritt auf Taiwan. *Zeitschrift für Ausländische Landwirtschaft* Special No.7, 92pp.

Webster, C.C.; Wilson, P.N. (1969) Agriculture in the tropics. London, UK; Longmanns 488pp. (Ed.3).

Wehrmann, J. (1970) [Fertilizer use in developing countries.] Düngung in Entwicklungsländern. Paper given at the 3rd Project Managers' Conference, Bonn.

Weitzenberg, H. (1962) [Water and land conservation in Africa. Suggestions based on experience on projects to combat drought within the framework of Technical aid—to secure and promote the social and economic development of African countries.] Wasser- und Boden-Erhaltung in Afrika. Auf Erfahrungen gestützte Vorschläge zu Projekten der Dürre-Bekämpfung im Rahmen der technischen Hilfe—zur Sicherung und Förderung der sozialen und wirtschaftlichen Entwicklung der afrikanischen Länder. *Schriftenreihe zum Handbuch der Entwicklungshilfe* No.6, 72pp.

Wilhelmy, H. (1974) [Rice cultivation and the food problem in Southeast Asia.] Reisanbau und Nahrungsspielraum in Südostasien. Kiel, German Federal Republic; Ferdinand Hirt 100pp.

Wrigley, G. (1971) Tropical agriculture. The development of production. London, UK; Faber 376pp.

Yaron, B.; Danfors, E.; Vaadia, Y. (*Eds*) (1973) Arid zone irrigation. Berlin; New York, USA; Springer Verlag 434pp. (Ecological Studies No.5).

Zörner, H. (1923) [Irrigation management in the light of teaching on farm management with particular reference to the German situation.] Die Bewässerungswirtschaft im Lichte der landwirtschaftlichen Betriebslehre unter besonderer Berücksichtigung der deutschen Verhältnisse. *Landwirtschaftliche Jahrbücher* 57, 605–665.

# Chapter 7

# Systems with Perennial Crops

Perennial crops are much more important in the humid tropics than they are in central and northern Europe. Table 25 provides a preliminary outline of the ecological and economic reasons why this is so.

## General Production Economics

The fourth major group of farming systems in the tropics is those cultivating tree and shrub crops. They are mostly found in the humid inner tropics. These include the uplands near the equator, the subhumid savannah and especially the humid and evergreen tropical rain forest. Peasant farmers in developing countries do not concentrate wholly on tree and shrub crops, but combine such crops with rain fed and irrigated farming, at least to an extent sufficient to meet their own food requirements. Fig. 31, illustrates a typical arrangement in the communal land of a Congolese village.

### Economic characteristics of perennial crops

In countries with tropical rain climates, a considerable proportion of total exports consists of products of perennial crops. Thus in 1975 exports from Indonesia were 41.6% rubber and 21.4% coffee; Ghana's exports were 98.1% cocoa, coffee and tea; Tanzania's 42.8% coffee and tea; Ethiopia's 33.2% coffee, tea and cocoa and those of Peru were 66.1% sugar. Plantation crops are of major importance to many other developing countries. Examples are coffee in Brazil, rubber in Malaysia, tea in India and Sri Lanka, sisal in Kenya, and sugar cane in Java and Cuba. Sisal and sugar cane fall half way between field crops and perennial crops. Sisal, sugar cane and tea tend to be grown on a large scale in plantations, so that the expensive factory plant can work at full capacity from a small collecting area. For sisal and sugar cane the additional need to transport a crop with so much bulky waste makes it imperative to grow the crop close to the factory. The fibre yield of sisal is only 3 to 4% but it is economically worth while to tansport the processed product all over the world. Almost the same applies to sugar.

On the other hand, oil palm, coconut palm and cocoa are true peasant crops. The wide variation in quality that follows from this is a disadvantage for the large amounts that are exported and often requires help from marketing boards. Rubber and coffee are crops which can be grown either by the large-scale farmer or by the peasant smallholder. Coffee is predominantly a plantation crop in Angola, Kenya, Tanzania and Zaire, while peasant farmer production has been developed in Uganda and Ethiopia. Because of the dry processing technique for coffee it is possible to produce even on the smallest area of land and with just a few trees.

The particular economic characteristics of tree and shrub crops can mainly be attributed to the fact that they are permanent crops. Table 25 shows their useful economic life varies from 3-5 years for pineapple to over 100 years for robusta coffee.

Table 25   Farm management characteristics of tropical tree and shrub crops.

| Tree or shrub crop | Ecological requirements | | | Economic characteristics | | | | | Expenditure and returns | | | Special remarks |
|---|---|---|---|---|---|---|---|---|---|---|---|---|
| | Height m/OD | Rainfall mm/year | Annual mean temperature °C | Peasant farm (B) or plantation (P) | Industrial processing | Irrigation | Economic life in years | Initial investment DM/ha | Labour input mh/ha/year | Natural yield dt/ha | Gross returns DM/ha | |
| **A. Tropical rainforest climate** | | | | | | | | | | | | |
| Cocoa | <600 | >1 300 | 24–25 | B,P | X* | – | >20 | 3 700 (Nigeria) | 300–1 000 | 4–8 | 1 300 (Nigeria) | 60% labour for 3 month harvest |
| Rubber | <600 | 1 500–2 500 | 25 | B,P | XX | – | 35 | 2 400 (Nigeria) | 960 | 3–5 (wild) 6–22 (grafted) | 1 300 (Nigeria) | can be grown on slopes |
| Oil Palm | <1 000 | 1 500–3 000 | 24–28 | B,P | X | – | 50 | 1 600 (Nigeria) | 350 | 6.0 | 850–1 700 | especially West Africa |
| Coconut Palm | Prefers coastal position | 2 000–2 500 | 25 | B | X | – | 80 | 2 500–8 000 | 560 | 11.7 | 600 | especially South East Asia |
| Pineapple | 100–500 | 1 000–1 500 | 21–28 | | XX | XX | 3–5 | . | 600 | 400–600 | 14 200 | fruit very difficult to transport |
| Abaca | <500 | 3 000–4 000 | 23–29 | B(P) | X | – | 0–15 | . | 850–1 200 | 12.0 (fibre) | 830 (fibre) | |
| **B. Humid savannah** | | | | | | | | | | | | |
| Robusta Coffee | 200–700 | >1 800 | 20–25 | B,P | X | X(XX) | >100 | 800 | 750 | 8–9 | 1 450 (Brazil) | must be grown within 150 km of a port |
| Plantain | <1 200 | 2 000 | 20 | B,P | – | XX (mostly) | 5–20 | c800 | 2 200 | 600 | 2 200 | typical subsistence fruit |
| Mango | <600 | >1 000 | 24–28 | B | – | XX | | | | | | |
| Sugar Cane | <600 | >1 200 | 25–28 | B,P | XX | . | 2–10 | 2 000–2 300 | 630 | 900–1 500 | 2 000–3 000 (Kenya) | 60% costs are for labour, 1968 average African yield 636 dt |
| Kapok | . | 1 100–1 850 | . | B | (X) | . | decades | . | . | 3.5–7.0 | | fibre for quilts and insulation |
| **C. tropical uplands climate** | | | | | | | | | | | | |
| Arabica Coffee | 600–1 800 | 900–2 300 | 17–20 | B,P | X | X(XX) | 30 | 1 500–3 000 (Kenya) | 1 600–2 600 | 5–10 | 1 500–2 300 (Kenya) | up to 90% harvest loss through CBD, spraying costs DM 700–800/ha |
| Tea | 2 000–2 400 | 1 500–2 400 800–1 000 | 20 | P | XX | . | >50 | 4 000–1 200 | 3 200–5 600 | 15.6 | 1 300–1 800 (Kenya) | |
| Castor oil | . | | . | | . | XX | several years | | . | 10–40 | | can be grown on slopes, seed contains 40–60% oil of high viscosity |
| Passion fruit | 1 600–2 500 | 1 050–1 500 | . | B | XX | . | 5 | 600–900 | 750 | 110 | 1 500 (Kenya) | juice used for drinks, ices, yoghurt, jam etc. |
| **D. Arid tropical climate** | | | | | | | | | | | | |
| Sisal | <1 800 | 750–1 300 | 16–30 | P | XX | . | 5–9 | 1 500–3 000 | 630–840 | 12–2 (fibre) | 1 200 (Tanzania) | only 3–4% of the crop is fibre |
| Citrus fruits | <700(2 000) | >1 200 | 17–20 | B,P | X | XX | 30–40 | 7 000 (Florida) | 1 300–1 600 | 300–380 | 10 200 | – |
| Date palm | ±0 | 75–500 | 25–32 | B(P) | X | X | 40–60 | . | 1 000 | 80 | 1 600 | to some extent a basic food stuff |

* X = desirable. XX = necessary.
Main Sources: Franke (1967); Piekenbrock (1958); Ruthenberg (1967); Ruhr-Stickstoff (1953–57); FAO (1970).

Fig. 31
Cropping system of a Congolese village. (Source: Dumont 1957, p.28)

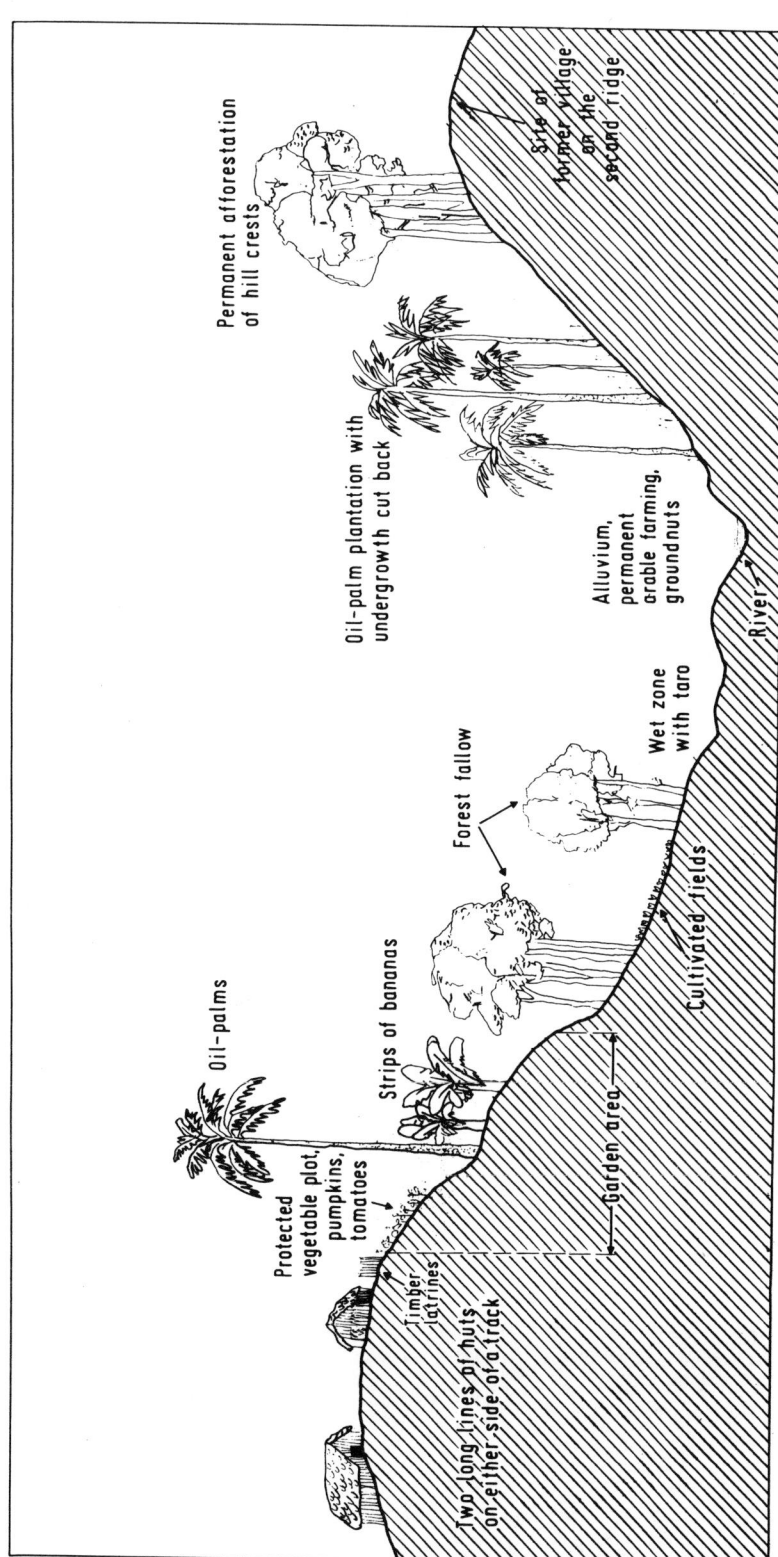

There are various consequences of these long cycles of production:

(1) A high initial investment is needed for planting and to cover the period when the young trees yield little or nothing. Costs are recouped as production gradually increases. For oil palm costs are recouped about four years after planting, for rubber after about six years and for cocoa after about eight years (Ruthenberg 1967).

(2) After the initial period yields are relatively stable because the deep root systems of established trees and shrubs make them fairly insensitive to annual and seasonal variations in rainfall.

(3) The constant ground cover which tree and shrub crops provide is similar to the natural forest vegetation of the humid tropics and thus maintains or at least greatly conserves the soil's fertility.

(4) There are harvesting problems because many types of tree and bush crop require high inputs of labour for harvesting and thus have high labour costs. These costs can only be absorbed when there are correspondingly high gross returns or low wages. Most tree and shrub crops must therefore be regarded as intensive crops.

(5) The minimum size of farm to provide a livelihood for an average peasant family is much less under perennial crops, or with irrigated farming, than with rainfed farming or extensive livestock.

## Comparison of the various systems

While these general agricultural and economic principles apply broadly to most perennial crops, there are quite wide differences between the very many crops in this group.

The selection of six tree crops in Table 26 illustrates the great difference between crops which help give the general system its flexibility. The initial investment for a particular crop is governed by the minimal-cost combination and thus appropriate changes in the production procedure are required as economic development proceeds. Even so, the relationship between the initial investment for the six crops listed is probably essentially peculiar to them. The longer the expected period of

Table 26   Important perennial crops.

| Economic characteristics (approx. values) | Tropical uplands | | | Tropical rain forest | | |
|---|---|---|---|---|---|---|
| | sisal | coffee | tea | oil palm | rubber | cocoa |
| Useful economic life in yrs. | 5–9 | 30 | over 50 | 50 | 35 | over 20 |
| Initial investment DM/ha | . | 1 350–3 000 | 4 000–12 000 | 1 600 | 2 400 | 3 700 |
| Labour hours per ha/yr | 630 | high | 3 200 | 350 | 960 | 300–2 000 |
| Gross income in DM/ha/yr | 1 200 | 1 500 | 1 300 | 465 | 1 300 | 1 300 |
| Crop processed in small plant (K), or factory (F) | F | K, (F) | F | (K), F | F | K |
| Specially suitable for peasant farms (B), or plantations (P) | P | B, P | (B), P | B | B, P | B |

Sources: Franke (1967); Ruhr-Stickstoff (1953–72); Ruthenberg (1967)

production, the more the farmer can afford to spend on laying out his plantation carefully and thoroughly.

The difference in labour requirements vary almost from one to ten as between oil-palm and tea, with the other crops lying between. This variation can basically be attributed to the different methods of harvesting and the different proportions of new planting each year. It illustrates how an appropriate tree or shrub crop can be selected to suit very different farming situations.

It is not, however, possible to deduce whether an individual crop is more suitable for peasant farming or for large-scale plantation farming simply from the amount of labour required. Much more important are the technical processing requirements of the harvested crops. While hand processing still produced a product of acceptable quality, tea was grown as a garden crop on small peasant holdings especially in Japan and China, simply because of its high labour requirements. Now that the world market demands, and is prepared to pay for, a better and more standard quality of tea, it must be processed under strict hygienic conditions in large factories. Tea has therefore quite clearly become a plantation crop in India, Pakistan, Sri Lanka and Java, in spite of the amount of labour it requires.

Sisal needs a relatively small amount of labour but the expensive plants required for processing must be worked to capacity from a small catchment area and the transport costs involved require large-scale operation. The oil-palm is a typical peasant crop in West Africa, although it requires relatively little labour. The low initial investment makes it popular and the processing problem has been solved by a changeover from home oil production to cooperative oil mills. Although its position is quite opposite, cocoa is also a peasant crop in West Africa. In this case the high initial investment costs are accepted because a great deal of labour can be employed productively on a small cropping area.

## A Comparison of Sisal and other Fibre Crops

Agricultural scientists have paid very little attention to the production of natural vegetable fibres other than cotton in the post-war years. Apart from cotton, developments in production of natural fibres are of minor importance for the industrialized countries, and in the developing countries agricultural scientists have been primarily concerned with the overwhelming problems of food shortages and hunger. As far as I can find out, no comparative study of the role of fibre-producing plants in the world economy has been carried out on a global scale. I will therefore try to break new ground by evaluating and comparing the ecological and economic conditions of production and the substitution, competition and complementary relationships of the world's various fibre plants, although only a few of them belong to the perennial crop group with which this chapter is mainly concerned.

### International division of labour

The world natural fibre industry is characterized by division of labour. This is necessary because, for economic and ecological reasons, the industrial countries cannot meet their needs from home production, while developing countries can produce surpluses for export. According to the FAO between 1964 and 1971 the combined use of cotton, wool, linen, silk and synthetic fibres increased from 13.7 to 16.2 kg per caput in developed countries, from 4.6 to 5.3 kg in countries with centrally planned economies and only from 2.6 to 2.8 kg per caput in developing nations. Over this time period, consumption increased by only 7.7% in developing countries, compared with 18.2% in the developed nations and by 1971, per caput consumption in developed countries was 5.8 times as large as that in developing countries. The industrialized countries cannot possibly meet this demand for textile fibres from their domestic production. Although flax is a plant which can be grown in temperate climates with long hours of daylight, its labour-intensive production requirements are not compatible with high levels of wages. Cotton, the most important textile crop, requires subtropical climatic conditions and therefore cannot be grown to any very great extent in the western industrialized countries, except in the southern part of the United States. Apart from hemp, fibres needed for use in sacking, upholstery, ropes, cables, string, nets, etc., are grown only in tropical climates. They are produced

Table 27   Exports of natural fibres from selected countries in 1970 as a percentage of national production.

| Country | Cotton | Sisal[1] | Country | Flax | Hemp |
|---|---|---|---|---|---|
| Angola | 79 | 96 | France | 232 | 170 |
| Brazil | 46[2] | 78 | Poland | 25 | 34 |
| Mexico | 46[2] | 27[3] | USSR | 9 | 4 |
| Tanzania | 38[2] | 120 | Hungary | 123 | 69 |

1) Including other agave plants, 2) 1971, 3) 1969

Source: FAO (1972)

almost exclusively in regions where there is little demand for them, so that, as Table 27 shows, the developing countries which grow these fibres have considerable surpluses for export. For many developing countries the balance of trade and, to a certain extent, the national economy are dependent on the export of natural fibres. For example, in 1970, cotton accounted for 44.6% of Egypt's exports by value; the figures for Sudan and Chad were 63.3% and 65.4%. In the same year, sisal exports amounted to 10.5% for Tanzania and for Pakistan jute provided 18.4% of the total value of exports.

Apart from ecological factors there are also economic considerations which would discourage the production of many varieties of fibre plant in developed countries. The cultivation of abaca, jute and sisal requires such a high input of labour that it would be impossible to start growing them in industrialized countries, even if it were ecologically feasible.

These economic and ecological requirements would not have been sufficient on their own to lead to the existing division of labour in the world fibre economy; technological considerations have also been important. In particular, the suitability of natural fibres for transport and their durability in storage create conditions which are economically favourable for the international exchange of goods. Thus, for most practical purposes the developing countries are the producers of fibre plants and the industrial nations are the consumers. A wide and varied stream of natural fibres of vegetable origin continually flows from the "lower" economic latitudes to the "higher".

**Major fibre crops**

*Production and countries of origin*

Table 28 summarizes information about the world's most important fibre crops. The cultivated area is geared to the production of only a few varieties of fibre. By far the largest area is under cotton (33.9 million ha), followed by jute (2.9 million ha), flax (1.5 million ha) and agave or sisal fibres (1.1 million ha). Between 1948/52 and 1972, cotton production increased 72%, sisal 75%, flax 38%, and jute and kenaf 72%. In the same period, production of hemp declined by 36% and abaca by 39%. In addition to being limited to few varieties of plant, as Table 28 shows, vegetable fibre production is also concentrated in specific countries. Three points deserve notice: (1) The ecological range for cultivation is especially small for jute, flax, fique and abaca or manila hemp, but is much greater for cotton and hemp; (2) flax and hemp, as temperate crops, are now mainly grown in Eastern European countries and (3) jute, sisal, fique, kapok and abaca are exclusively tropical plants. With the exception of jute they are perennial bush or tree crops. Production is therefore inelastic in the face of fluctuations in world market prices.

*Processes of production and utilization of fibre*

The production of most fibre crops is in fact, inelastic because of the need to construct expensive works to extract the fibre in the growing areas. For jute, the yield of fibre is between 4.5% and 7.5% of the harvested material; for sisal and abaca, the proportion is between 2% and 5%. This means that for the typical tropical fibre crop which has these very large residues and which is grown in areas with extremely severe transport problems, fibre extraction must take place on, or very near to, the holdings where the crop is grown. However, even in developing countries only jute, hemp and abaca are suitable for small scale processing and it is usually necessary to invest in expensive factory installations. The expenditure on such equipment must be recouped before the production programme can be changed without economic loss. For example, sisal plantations require an initial investment of DM2750 per ha., this includes rail transport from the fields and the cost of the factory. To make full use of a factory of an economic size, some 1200 to 2000 ha of sisal must be grown in the immediate vicinity of the factory (Ruthenberg 1967 p.187, Franke 1967 p.364). The reason for the tendency towards large-scale plantations and the "Agave landscape" around the factory is the concentration of the area under cultivation to economize on transport costs. Once the 2–5% of fibre has been separated from the leaves, washed, dried, brushed and baled, transport costs become negligible. It thus becomes possible to use the sisal produced in the equatorial zone for manufacturing twine and rope in northern Europe.

Natural fibres sent from developing countries to the industrialized west are normally exported in the form of bales of raw fibre. The manufacture into tow, ropes, twine, textiles, etc., is carried out more satisfactorily in the industrial countries than in the developing countries, because further transport costs are no longer significant and the processes are capital-intensive in nature rather than labour-intensive.

Cotton has a fibre yield of 30% to 40% and consequently cotton cultivation is more flexible. It is economic to transport the harvested crop over greater distances to the cotton gin and baling installations than is possible for other fibres. It is eminently feasible to grow cotton on smallholdings and process it in co-operative plants. In the

Table 28   Major fibre crops: world production, countries of origin, processing and utilization.

| Crop | Area in 1972 ha × 10² | Yield kg/ha | Fibre Production t × 10³ p.a. fibre | | | Major producers | Processes to obtain fibre | Uses |
|---|---|---|---|---|---|---|---|---|
| | | | 1948/52 | 1961/65 | 1972 | | | |
| Cotton | 33 909 | 3.8 | 7 581 | 10 798 | 13 031 | USA 23% USSR 19% China 11% | cotton gin bale press | yarn, thread, knitted and woven textile upholstery |
| Flax | 1 536 | 4.1 | 463 | 679 | 636 | USSR 70% Poland 9% France 7% | retting, rinsing, crushing, drying, beating, swingling, combing | bed and table linen, towels, work clothes, awnings, sailcloth, thread |
| Hemp | 423 | 6.0 | 397 | 342 | 254 | USSR 32% India 18% China 10% | only 'technical' fibre is processed | ropes, tow, nets |
| Jute and kenaf | 2 881 | 13.2 (Jute) | 2 025 | 3 336 | 3 486 | Bangladesh 35% India 31% China 15% | crushing, soaking to soften fibre, extracting fibre, bleaching, drying | sacks, upholstery, fabric coverings, ropes, tow |
| Sisal (and other agave fibres) | 1 054 | 7.1 (Sisal) | 438 | 830 | 766 | Brazil 27% Tanzania 20% Mexico 19% | retting, removal of fibre from leaf, washing, drying, brushing, baling | twine, cord for baling, ropes, tow, carpets, upholstery (inferior to abaca for ships' ropes) |
| Abaca (Manila hemp) | 157 | 4.8 | 124 | 112 | 76 | Philippines 89% Malaysia 5% Ecuador 4% | removal of fibre strips from leaf, sorting, scraping, drying, sorting, bundling | ships' ropes, nets, ropes, hammocks, furniture coverings, string, cable ropes, transmission belts |

Sources: FAO (1973); Franke (1967)

irrigated Gezira area along the Blue Nile in the Sudan, the syndicate takes over the de-seeding, storage and utilization of the cotton produced on small family holdings. Since cotton is an annual crop it is very easy to expand or contract production. Table 28 provides further information about the production and utilization of other fibre crops over the world as a whole.

### Diverse ecological requirements

Although this examination is limited to the six most important fibre producing crops in the world economy, they cover a wide range of different locations because of their ecological requirements. The physiologically optimum conditions are shown in Table 29. Briefly flax prefers temperate climates, hemp is suited to temperate and sub-tropical climates, and cotton to sub-tropical regions and the arid areas bordering the equatorial zone. Sisal grows in tropical highland areas and jute and abaca in tropical rain forest and humid savannah climates.

## Economic assessment of fibre crops in crop rotations

Cotton resembles maize and sugar cane in that it is one of the few crops in the world that can be grown under conditions of monoculture for decades. Thus in the Mississippi delta cotton is considered the best crop both to precede and to follow the cultivation of cotton. Crop rotation only occurs when the soils on a farm are unsuitable for maize or cotton—if the land is too damp for cotton or too dry for growing maize as grain. Examples are shown in Table 30, sections C and D. Crop rotation is the rule in the southwestern regions of the United States. The soil is more uniform and it is possible to regulate the amount of water to provide the optimum amount for cotton on land that can be irrigated (Table 30, A). Among other plants, barley and lucerne are grown along with cotton; barley, in contrast to cotton, needs winter irrigation while lucerne will give six to eight cuttings under semi-arid conditions with irrigation. In the Sudanese Gezira, the cultivation area of each farm is divided into one third for cotton, one third for subsistence crops and one third fallow.

In the rice and tea growing areas of China, crop rotation is on a two year pattern; cotton is the regular summer crop but the winter crop alternates between wheat, barley or rape in the first winter and legumes in the second, the legumes providing humus for the soil (Franke 1967 p.316). Cotton is grown as a cash crop on tropical smallholdings in rotation with subsistence crops such as groundnuts, soya, sorghum, maize, wheat, sesame, manioc and yam as well as rice and sugar cane (Franke 1967 p.315). (See also Table 30, E and F.)

Hemp can be grown continuously on the same land for many years without any noticeable decline in yield; Italy is an example. The considerable tolerance of hemp is illustrated in Table 31, which shows that as a rule hemp is grown in alternate years. In addition, hemp is considered ideal for growing before any other crop because it rids the land of weeds and prepares the soil for the next crop. Hemp requires very little care owing to its vigour, which makes it resistant to being smothered by weeds; it has also produced good results as a pioneering crop on marshy ground. On the other hand, flax will not tolerate being grown on the same land for two years running and, as Table 32 shows, the crop rotation should allow seven year intervals between plantings.

Table 29   Ecological requirements of the major fibre crops of the world.

| Ecological requirements | Cotton | Jute | Flax | Sisal | Abaca |
|---|---|---|---|---|---|
| Climatic zone: | | | | | |
| most favourable | subtropic | tropical rain forest | temperate maritime | tropical highland areas | tropical rain forest |
| favourable | semiarid zones | humid savannah | temperate continental | tropical arid areas | humid savannah |
| Temperature requirements: | | | | | |
| range, °C | 15–40 | 15–40 | temperate | 16–32 | 23–29 |
| optimum, °C | 26–28 | 24–35 | warm summers | | 27 |
| Precipitation, mm/annum | | | frequent rains | | |
| range | 500–1500 | 900–2500 | especially in May–June | 250–1800 | 2500–4000 |
| optimal | 800–1000 | 1500–2000 | | 1000–1300 | 3000–4000 |
| Altitude, metres above sea level | | | | | |
| range | up to 1500 | up to 30 | . | up to 3200 | up to 500 |
| optimal | lowland | lowland | lowland | up to 1800 | lowland |
| Soil requirements | light soil, with good drainage | sandy loam to loamy clay, esp. alluvium | optimal is light loam | light, porous, neutral soil | fertile, deep, light soil |
| Other ecological requirements | >180–200 frost free days. Likes warmth and light | sensitive to duration of daylight. 70–90% humidity | wet climate, high atmospheric humidity | sensitive to frost. Daily temperature variation <7–10°C | dry periods can cause damage |

Source: Franke et al (1967).

Table 30    Examples of cotton in crop rotations.

1. Arid climates (cultivation with irrigation)

A

Southwest U.S.A.
1–3. lucerne
4–5. cotton
6–7. summer barley

B

Nile Delta
1. cotton (February to October)
2. clover, beans, vegetables (November to May)—fallow
3. wheat, summer barley (October to May) maize (July to October) — clover as catch-crop (November to December)

2. Subtropical areas with warm summers (no irrigation)

C

Mississippi Valley
1. cotton
2. clover grass
3. corn (maize)
4. winter wheat

D

1. cotton
2. corn (maize)
3. oats

3. Tropics with seasonal rains[1]

E

non-irrigated cultivation, 1200–1400 mm of rain p.a.

Uganda
1–2. fallow
3. cotton
4. millet
5. manioc, sweet potatoes

F

irrigated cultivation, 750–850 mm of rain p.a.

Uttar Pradesh, India
1. sugarcane
2. maize, wheat, chickpeas
3. cotton, fodder

1) In India cotton is grown in conjunction with Italian millet, coriander and mountain rice. Monoculture does not occur.
Source: H. Ruthenberg (1967) pp. 131 ff and 159.

Table 31    Examples of hemp in crop rotations.

| A | B | C |
|---|---|---|
| Central Russian Forest steppe | Lower plains of the River Po | Campania, Italy |
| 1. hemp | 1–3. lucerne | 1. hemp |
| 2. sugar beet | 4. hemp | 2. corn or beans followed |
| 3. hemp | 5. turnips, tobacco | by legumes as green |
| 4. potatoes | 6. hemp | manure |
| 5. hemp | | |
| 6. field beans | | |

Source: Könnecke, 1967, p. 303.

Table 32   Examples of flax in crop rotations.

| A<br>Mixed forest zone<br>Grodno district USSR | B<br>Moscow district | C<br>Northern Japan |
|---|---|---|
| 1. flax | 1–2. clover | 1. flax |
| 2. potatoes | 3. flax | 2. winter wheat |
| 3. silage maize | 4. semi-fallow | 3. soya beans |
| 4. field fodder | 5. winter grain | 4. summer barley |
| 5. winter grain | 6. potatoes | 5. maize for grain |
| 6. field fodder | 7. oats | 6–7. summer grain |
| | | 8. clover |
| | | 9. potatoes |
| | | 10. oats |

Source: Könnecke, 1967, p. 302.

In order to utilize a flax racking machine fully, it is essential to have very large flax growing enterprises or to use the machine for more than one farm. Fodder crops that take over a year to mature are particularly suitable for cultivation in the year before flax (Table 32, B). However, potatoes, sugarbeet, leguminous crops and annual fodder plants are also grown. Because of the demands that flax makes on soil fertility, it is desirable to have a complete crop rotation prior to that which includes flax. On the other hand, flax fits in well with rotations of leafy crops and possesses useful qualities both as a preparation for another crop and, to a certain extent, as a cover crop.

*Economics of labour input*

All six fibre crops listed in Table 33 are labour-intensive. Labour accounts for 68% of total production costs in the case of sisal and 74–80% for jute. Because of the considerable size of factory installations and the extensive transport system needed, sisal must be grown on large holdings. This is economically feasible because the work of new planting, weed control and harvesting is spread evenly throughout the year (Ruthenberg 1967 p.187). Sisal is more amenable to mechanized cultivation than most perennial crops, provided that the size of plantation is sufficient to cover the overhead capital costs for weed control and transport of the harvested leaves. Some weeds are allowed to grow between the rows of sisal plants to control soil erosion. The absence of any cultivation over a period of years has a beneficial effect on the fertility of tropical soils. A modern decorticator has the annual capacity to produce 1500 tonnes of fibre. This output requires the transport of 40 000 tonnes of leafy material within the plantation. To provide this transport and meet the daily needs of the decorticator it is necessary to have three 50 HP diesel engines and at least 60 rail trucks (Franke 1967 p.365). Despite this the actual harvesting of the leaves is still done by hand.

In the case of the other five fibre crops the high labour requirement is aggravated by the marked peak in the demand for labour at harvest time. These two problems are much more easy to solve on small farms than on large scale enterprises. Large plantations are only viable if they can use technical innovations to make drastic reductions in the demand for labour and if it is economically viable to mechanize, and spread labour requirements more evenly over the whole season.

Table 33   Management characteristics of the production of fibre crops for world markets.

| Management characteristics | Cotton | Cotton | Flax | Jute | Hemp | Hemp | Sisal | Abaca |
|---|---|---|---|---|---|---|---|---|
| | Industrial countries | Developing countries | Industrial countries | Developing countries | Industrial countries | Developing countries | Developing countries | Developing countries |
| Favoured size of holding[1] | SL | S(L) | SL | S | SL | S | (S)L | S(L) |
| Growing period, in months | 7–8[2] | 6–7[2] | 3–4 | 4–5 | 3–4 | 3–4 | 8–10 years[3] | 10–15 years[3] |
| **Labour economics:** | | | | | | | | |
| Man hours per ha/annum. | 124–200 | 1 450–1 550 | 210 | 1 200 | 80–90 | . | 630–840 | 850–1 200 |
| Harvest as % labour | high | peak at harvest | c. 80 | 28–32 | . | . | 41–46[4] | 17–32 |
| Length of harvest, months | up to 2.5 | up to 2.5 | 0.5 | 1–2 | 0.5 | 0.5 | 12 | every 3–8 months |
| Harvest method | machine | hand | rack machinery | hand | combine harvester | machete, sickle, etc. | hand | hand |
| Fibre as % harvested material | 30–40 | 30–40 | 20–24 | 4.5–7.5 | 18–25 | 18–25 | 2–5 | 2–4 |
| Fibre extraction[5] | F | F | F | C.I. F | F | C.I. | F | C.I. (F) |
| Other characteristics: | Hand picking requires 3–4 pickings. Loss due to pests, diseases and weeds approximately 50% of world production. | | Interval between crops >7 years | 70–80% of production costs are labour costs. Plants grow to height of 4–5 m. | Production costs have risen sharply in Western Europe | Deters weeds; this is very important in developing countries | Approximately 68% of production costs are labour costs. Fibre extraction <2 days after cutting | Rational use of machinery would require units of 1 000 to 2 000 ha |

1) S = Family farm, L = Large enterprise   2) 2 years if necessary   3) Economically useful life of plant
4) Including rail transport from field to factory   5) F = Factory, C.I. = Cottage Industry
Sources: Piekenbrok (1958) and others.

This solution is hardly possible to apply to the growing of abaca, which is produced almost exclusively on smallholdings, Franke (1967) points out that before the second world war, almost half the Philippine holdings growing abaca were smaller than 1.2 ha and that rational mechanization of fibre production requires holdings of between 1000 and 2000 ha. This is a requirement which is totally incompatible with conditions in overpopulated developing countries. In the case of jute one third of the extremely large labour input of 1200 manhours per ha is concentrated into the 1–2 month harvest period. In effect, this limits production to minute areas of land on family holdings which are already short of land. The classic centre of jute production is Bengal, where the average size of holding is only 1.26 ha. Of this land, 88.1% is used for rice, 5.1% for other subsistence crops and only 6.8% for jute. It has been calculated that an individual farmer produces only 7–9 bales of jute annually, working with the same tools and methods as his forefathers would have used a century ago (Malik 1966 p.47 ff.). Jute is an extremely labour-intensive and capital-extensive crop.

On the other hand, mechanization of harvesting has been very successful in the case of cotton where picking machines are used after the crop has been sprayed with defoliation agents. Racking machines can be used to mechanize flax harvesting and hemp can be harvested using special combine harvesters. As a result, it is an economically viable proposition to grow any of these three fibres on large-scale enterprises in industrial countries. Full mechanization of tillage, cultivation, and harvesting has reduced the labour requirements for growing cotton to one-tenth of the level before mechanization. Without mechanization, the USA which is the world's second largest producer, would find it economically impossible to grow cotton. There is a wide range of ways in which capital investment in machinery can be combined with the use of manual labour in the production of cotton. This makes it possible to grow cotton on farms of very different sizes in countries at all stages of economic development, as inputs can be combined in many different ways to minimize production costs. The economic feasibility of growing cotton has therefore considerably widened. On the other hand the introduction of picking machines has narrowed the ecological limits for growing cotton. Mechanized production is concentrated on irrigated land in dry areas, as mechanical harvesting is more dependant on dry weather than hand picking.

*Comparative productivity*

Table 34 compares the productivity of the six fibre crops. The validity of the comparisons depends on the assumptions made in deriving the estimates and the table should be interpreted only as giving orders of magnitude. Productivity varies considerably because inputs of labour, yields and prices differ from one country to another, between holdings, and from one season to another. It was only possible to estimate gross productivity, as it is not easy to quantify the inputs. This makes productivity look much higher in the industrial countries, because their inputs of capital are so much higher than in developing countries. Fair comparison is therefore only possible among industrial countries or among developing countries and not between the two groups.

For the developing countries, especially those that are densely populated, the most suitable yardstick for comparison is land productivity (returns per ha). Jute and abaca can be grown under similar conditions and make comparable demands on labour. Jute is shown to be clearly superior to abaca and this explains the 72% expansion in jute

Systems with perennial crops   137

Table 34   Comparison of the productivity of world fibre crops.

| Characteristic | Unit | Flax, Belgium | Hemp, Italy | Cotton, USA | Sisal, Tanzania | Jute, Bangladesh | Abaca, Philippines |
|---|---|---|---|---|---|---|---|
| Population density | – | high | high | low | low | high | high |
| Stage of development | – | industrial | industrial | industrial | developing | developing | developing |
| Climatic zone | – | maritime, temperate | sub-tropical | warm summers | tropical, seasonal rain | wet tropics | wet tropics |
| Annual precipitation | mm | 600–1 000 | 700–800 | 250 | 1 000–1 300 | 1 500–1 700 | 3 000–4 000 |
| Altitude | metres above sea level | lowland | lowland | 200 | 1 000–2 000 | up to 30 | up to 500 |
| Size of holding | ha | 10 | 20 | 130 | 1 200 | 2.4 | 1.2 |
| Labour input | man-hours per ha/p.a. | 210[1] | 82[2] | 124[3] | 630[4] | 1 200 | 1 000 |
| Yields | 100 kg/ha | | | | | | |
| Fibre | 100 kg/ha | 9.5 | 10.0 | 4.9 | 20.0 | 12.0 | 12.0 |
| Seed | 100 kg/ha | 6.5 | 7.0 | 8.6 | – | – | – |
| Production costs | | | | | | | |
| Fibre | DM/100 kg | 208 | 213 | 150[5] | . | 88 | 69 |
| Seed | DM/100 kg | 34.30 | 35.00 | 14.80 | . | – | – |
| Gross income | | | | | | | |
| Fibre | DM/100 kg | 1 976 | 2 130 | 735[5] | 1 200 | 1 056 | 828 |
| Seed | DM/100 kg | 206 | 245 | 127 | – | – | – |
| Gross productivity | | | | | | | |
| Returns on land | DM/ha | 2 182 | 2 375 | 862[6] | 1 200 | 1 056 | 828 |
| Labour productivity | DM/man-hour | 10.40 | 29.00 | 695[6] | 1.90 | 0.88 | 0.83 |
| Farm workers' wages | DM/man-hour | 4.50 | 3.25 | 5.00 | 0.49 | 0.17 | 0.25 |

1) Racking machine   2) Special combine harvester   3) Irrigation, defoliation, picking machine
4) Including rail transport from field to factory and decortication   5) Decreased by 46–62%, 1955/63 to 1971   6) Decreased by 40–56%, 1955/63 to 1971
**Sources:** Franke et al (1967); Pössinger (1967); Ruthenberg (1967).

production (including kenaf) between 1948/52 and 1972. Over the same period, the production of abaca or manila hemp fell by 39%.

After the two thirds fall in world market prices for sisal during the last two decades productivity of land growing sisal is DM1200 per ha and these returns make it one of the less profitable tropical tree and shrub crops. By comparison, cocoa and rubber producing areas in Nigeria yield returns of DM1300 per ha, coffee in Brazil DM1450, tea in Ceylon approximately DM1600 and coffee in Tanzania DM2200. Coffee and tea are quality products that need very much higher inputs of labour than sisal, so that sisal has considerably better returns per man hour of labour. With present technology, sisal growers would find a rise in wages more practicable economically than tea or coffee growers.

In industrial countries with high wage levels, labour productivity is now much more important than land productivity. In the three industrial countries in Table 34, productivity per man hour for hemp, flax and cotton has markedly declined. This is all

Table 35   World production of major textile fibres.

| Type of fibre | 1948[1]–1952[1] | | 1960 | | 1971 | |
|---|---|---|---|---|---|---|
| | t × 10³ | % | t × 10³ | % | t × 10³ | % |
| Natural fibres | | | | | | |
| Pure wool | 1053 | 9.7 | 1510 | 9.2 | 1587 | 6.8 |
| Flax fibre | 463 | 4.3 | 650[2] | 4.0 | 648 | 2.8 |
| Cotton | 7600 | 70.5 | 11 000 | 66.7 | 11 799 | 50.9 |
| Artificial fibres[1] | | | | | | |
| Cellulose fibres | 1560 | 14.4 | 2603 | 15.8 | 3460 | 14.9 |
| Synthetic fibres | 115 | 1.1 | 710 | 4.3 | 5705 | 24.6 |

1) Artificial fibres for 1948–1954. 2) Estimated value.
Sources: FAO (1971), UNO (1963 and 1972).

the more serious as the mechanization of harvesting becomes more expensive. The gross returns on labour must also cover all expenditure and interest on capital before net income can be assessed. As these gross returns become very close to the actual level of wages paid to farm workers, it becomes doubtful whether it is still an economic proposition to grow flax and cotton in industrial countries. Despite this, world fibre production rose between 1948/52 and 1972. For flax the increase was 38% and for cotton 72%. This was possible during a period when there was pressure for higher wages because of a considerable increase in yields and lowering of fixed costs per unit of product and because of a shift in the location of production to East European countries and, in the case of cotton, to developing countries. In 1957, the USA still produced over half the world's total production of cotton but by 1972 their share had fallen to 23%. Cotton lost much ground in the old US cotton belt and production is now more concentrated in the south-western states, where large enterprises, dry weather for harvesting and level ground allow full mechanization, and where irrigation produces record yields per ha.

The productivity quotient (land productivity divided by labour productivity) suggests that under present conditions flax, hemp and cotton are more suitable for growing in industrially orientated countries with high labour productivity, whereas sisal, jute and abaca which need high land productivity are more suited to the developing countries. The productivity quotient of course varies widely as capital is substituted for labour, but it is primarily determined by the characteristics of the different fibre crops. The distribution of fibre-producing plants in different climatic zones is thus partly dependent on ecological conditions but is also rational in economic terms. Tropical developing countries supply fibre with high land productivity, while fibres supplied by industrial countries with subtropical and temperate climates are predominantly those with a high labour productivity.

The production of fibre crops is highly sensitive to fluctuations in world market prices. The economic risks in the case of cotton, jute and sisal are considerable, because fluctuations in product prices do not affect production costs and as a result the losses or gains directly affect the individual farmer.

Table 36    World consumption of major textile fibres.

|  | 1961/65 | | 1969 | | 1972[1] | |
|---|---|---|---|---|---|---|
|  | total[2] | per caput[3] | total[2] | per caput[3] | total[2] | per caput[3] |
| Cotton | 10 303 | 3.2 | 11 555 | 3.2 | 12 094 | 3.1 |
| Wool | 1 545 | 0.5 | 1 699 | 0.5 | 1 735 | 0.4 |
| Flax | 677 | 0.2 | 783 | 0.2 | 765 | 0.2 |
| Silk | 33 | – | 39 | – | 41 | – |
| Cellulose fibres | 3 049 | 1.0 | 3 485 | 0.9 | 3 475 | 0.9 |
| Synthetic fibres | 1 394 | 0.4 | 4 380 | 1.2 | 6 445 | 1.7 |
| Total | 17 001 | 5.3 | 21 940 | 6.0 | 24 555 | 6.3 |
| World population, millions | 3 230 | | 3 647 | | 3 875 | |
| Percentage of consumption | | | | | | |
| Natural fibres | 74 | | 64 | | 60 | |
| Artificial fibres | 26 | | 36 | | 40 | |

1) Estimated value. 2) t $\times$ 10³, 3) kg.
Source: FAO.

## Competition between natural and artificial fibres

### Expansion of artificial fibres

As Table 35 shows, artificial fibres have increased their share of the world's output of major fibres from 15.5% in 1948/52 to 39.5% in 1972. This increase was almost entirely due to the expanded production of synthetic fibres; production of cellulose fibres was almost stagnant. This relative decline from 84.5% to 60.5% in natural fibre production does not represent a decline in absolute terms. During this same period because of greatly increased demand the output of wool rose from 1.1 to 1.6 million tonnes, flax fibre from 0.5 to 0.6 and cotton from 7.6 to 11.8 million tonnes. As Table 36 shows, consumption of natural textile fibres per caput hardly changed between 1961/65 and 1972 but total demand rose in proportion to the increase in world population. For the individual consumer, artificial fibres have not replaced natural fibres but have supplemented them. The increase in per caput consumption of textile fibres from 5.3 to 6.3 kg per annum is almost identical to the growth in per caput consumption of artificial fibres.

### The slump in prices for natural fibres

This assessment does not mean that the development of artificial fibres has had no effect on the production of natural fibres or that a new consumer commodity has entered the market without causing competition. Natural and artificial fibres are substitutes and not

Table 37   Sensitivity of fibre crop production to the slump in prices.

| Item | Unit | 1948/52[1] | 1961/65 | 1971 | 1971 as % 1948/52 |
|---|---|---|---|---|---|
| **Cotton** | | | | | |
| U.S.A. producer prices | US cents per kg | 83 | 68 | 63 | 76 |
| World fibre yield | 100 kg per ha | 2.4 | 3.3 | 3.6 | 150 |
| World cultivated area | million ha | 31.0 | 33.0 | 33.1 | 107 |
| **Sisal and other fibre agaves** | | | | | |
| UK import prices | US cents per kg | 49 | 31 | 18 | 37 |
| World cultivated area | 1000 ha | 703 | 1106 | 1059 | 150 |
| **Jute** | | | | | |
| Pakistan export price | US cents per kg | 80 | 31 | 29[2] | 36[2] |
| World fibre yield | 100 kg per ha | 11.8 | 14.0 | 13.5 | 114 |
| World cultivated area | 1000 ha | 1717 | 2560 | 1859 | 108 |
| **Abaca** | | | | | |
| Philippine wholesale price | US cents per kg | 41 | 29 | 43 | 105 |
| World cultivated area | 1000 ha | 302 | 200 | 152 | 50 |

1) Prices from 1950 to 1952.   2) 1970.
Sources: FAO (1965, 1971, 1972)

complementary consumer goods. Within a wide field the consumer can choose which fibre he uses. The decision in favour of natural or artificial fibres will depend on factors like price or quality. Thus technical progress in the manufacture of artificial fibres which improved their quality and reduced their price pushed down the prices of natural fibres to almost ruinous levels. As Table 37 shows, cotton prices fell by 24% between 1948/52 and 1971, sisal prices by 63% and jute prices by 64%. This happened even although consumption per caput of natural textile fibres remained constant and world production rose considerably. The nature of the cause and effect can be inferred. Fierce competition from artificial fibres made it necessary for natural fibres to lower their prices if they were to maintain or even increase their sales.

*Adjustment in the cultivation of fibre crops*

Producers of natural fibres adopted various strategies to adjust to this drastic fall in prices some of which Table 37 helps to illustrate. Between 1948/52 and 1971, cotton yields rose by 50% and full mechanization of harvesting became possible on large enterprises. Both these factors reduced production costs and made it possible to absorb the pressure on prices, thus the 24% fall in price coincided with an increase of 7% in the cultivated area. At the present stage of technology industrial countries have carried rationalization as far as is possible. As a result any further fall in world prices, coupled with a rise in wages and in the income expectations of producers, would lead to a marked shift in the location of cotton production to developing countries.

Abaca production has responded to falling prices in a different way. The high price of capital goods in the developing countries of the equatorial zone made it impossible to adopt yield increasing inputs or labour-saving equipment on a wide scale. Mechanization was in any case made difficult because of the small size of production units. As a result, abaca producers had no way of reducing costs to combat the trend in world prices and indeed despite a small price rise the cultivated area declined by 50%.

Producers of sisal reacted to the 60% slump in prices by increasing the area grown by 50%. This was possible because large sisal plantations were able to take rapid advantages of mechanical and technical progress which led to a dramatic fall in production costs. According to Ruthenberg (1967) labour requirements for sisal production in Tanzania in 1965 had fallen to 35 workers per ha, compared with 70–100 in 1955.

The area under jute expanded slightly (8%) between 1948/52 and 1971, although Pakistan's export prices dropped by 64%. There has been very little increase in output per ha and methods of cultivation have not changed very much. This tolerance of such a drastic reduction in income can be explained by the fact that at an advanced stage of growth jute will tolerate being flooded. This makes it possible to combine the cultivation of jute and rice. In this situation, increased yields of rice as a subsistence crop have made it possible for smallholdings in areas like the river lowlands of Bengal and Amazonia to increase production of cash crops, particularly of jute. The high labour requirement for jute is unimportant in such areas where there is permanent rural underemployment.

**Summary on fibre crops**

The production of natural fibres occurs within a framework of worldwide division of labour, made possible by the properties of the fibres, i.e., ease of transport and stability of the fibres, and the more or less strictly determined ecological conditions necessary for the growth of the more important fibres.

Of the important fibre plants flax and hemp are grown in the temperate climatic zone; cotton and sisal require the dry tropical and highland tropical climates, while the wet tropics produce jute and abaca. Because the fibres have different properties, they are only partly interchangeable, with the result that world trade between the different climatic zones is necessary.

The division of labour in fibre production is determined not only by ecological conditions but also by economic factors. Jute and abaca are very labour intensive and require a high level of soil productivity; the cultivation of these crops is therefore eminently suitable for overpopulated developing countries. Flax and hemp are characterized by a high level of labour productivity and can remain relatively economical for some time in a situation of rising wages. Cotton and sisal occupy a midway position between the other two groups. Technically, these two crops are suitable for variable combinations of labour and capital in the production process and are therefore able to adapt to changing economic conditions.

Natural fibre production is under strong pressure from the price competition of artificial fibres. Abaca production has declined because price changes could not be matched by lower unit costs of production. Although there was a big fall in the price of cotton, this was compensated for by an expansion in cultivated area and lower production costs resulting from irrigation, fertilizer use, pest control and mechanization, often only made possible by changes in the location of production. Jute cultivation has expanded slightly, despite a sharp fall in price. In this case, expansion can be attributed to the location of production and the methods used. A similar slump in price has occurred with sisal, but with the economies of scale in production on large plantations the cultivated area increased by 50%.

During the last decade, total consumption of natural textile fibres per head of population has remained almost constant in absolute terms, so that production could be expanded in proportion to the rise in world population. The introduction of artificial fibres led only to a relative reduction in the consumption of natural fibres. Between 1961/63 and 1972, the share of natural fibres in world consumption fell from 74% to 60%. The relative expansion of artificial fibres from 26% to 40% in the same time span has not been at the expense of natural fibre consumption but arises from the over-all-increase in world per caput consumption of textile fibres, from 5.3 kg in 1961/65 to 6.3 kg in 1972.

## Coffee, Tea and Cocoa Production

### Production and consumption

All three of these crops supply a harvested product which has a high value per unit of weight, is easy to pack, and store and therefore easy to transport. Coffee, tea and cocoa are semi-luxury commodities in high demand, greatly sought after in industrial countries where consumption is highest and where production is not possible. On the other hand consumption of such semi-luxuries is low in the tropical areas where they are produced and where consumers have very limited purchasing power. This has long led to a division of labour with vast quantities of tea, coffee, and cocoa flowing continuously from the developing producer countries to the industrial consumer countries. World exports of tea, cocoa and coffee in 1976 were in a ratio 1:2.5:7.9 to each other and had a total value of US $11 960 000 million. Table 38 shows how consumption increases in the course of development without taking account of production possibilities.

In 1957 India consumed only 31% of its total tea production, Sri Lanka only 4%, China 28% and Indonesia 14% (Franke 1967). In 1976 world production of tea was 1.63 million tonnes, of coffee 3.65 million tonnes and of cocoa 1.39 million tonnes. The geographical concentration of production is quite exceptional. Thus 31.4% of tea is produced in India and 19.9% in China while 14.0% of coffee is grown in Brazil and 23.0% of cocoa in Ghana.

*Climatic conditions*

The reasons for this concentration of production in certain areas are less economic than ecological. The breakdown by the main climatic zones is as follows: (1) In the sub-tropics tea is practically without competition, only in a few exceptional positions is it also possible to grow arabica coffee. (2) The tropical highlands provide suitable conditions for both tea and arabica coffee which are both important in these areas. (3) In the humid tropical lowlands robusta coffee, cocoa and the cola tree are all competi-

Table 38    Per caput consumption of coffee and tea.

| Product | Year | Producing developing countries | | Consuming industrial countries | |
|---|---|---|---|---|---|
| Green coffee (equivalent) | 1975 | Indonesia[1] | 0.14 | USA | 5.73 |
| | | Mexico[1] | 0.5 | Denmark | 13.27 |
| | | Brazil[1] | 3.0–4.0 | Sweden | 13.54 |
| Tea | 1973/1975 | Sri Lanka | 1.5 | New Zealand | 2.5 |
| | | Morocco | 0.8 | England | 3.5 |

1) roasted coffee 1958/60.
Sources: Franke et al (1967); UNO (1967).

tive crops. Coffee does better in the areas further from the equator; nearer to it the other two crops have the advantage. Given these ecological relationships between tea and arabica coffee and between cocoa and robusta coffee the extent to which they are economic competitors for specific locations depends very much on management factors.

## Peasant farms or large scale plantations?

This choice depends very much on whether the harvested product is suitable for small scale processing or whether it requires to be handled in technically advanced large factories. This is affected both by the nature of the crop and by the intended end use, either on the domestic market or as an export commodity. In general development is moving away from the small scale plant towards the large factory. One reason for this is the pressure towards mechanization caused by higher wages and the fact that mechanization is easier to introduce at the processing stage than during harvesting. The second reason is the increasing demand for a high quality product in the industrial countries and the fact that this is hard to achieve with small scale peasant processing plant. Such developments favour large plantations while on the other hand many settlement schemes in the new nation states are breaking up the old European plantations. The question of peasant farms or plantations is thus one of conflicting pressures in both directions.

As Table 39 shows all three processing methods are exceedingly complicated. Even so it is possible to grow cocoa on peasant farms as the different processing activities can be carried out at a co-operative level. The progress made by Ghana, the world's major cocoa producer, over the last 20 years is primarily due to the successful operation of the Cocoa Marketing Board, which works mainly on a co-operative basis. Cocoa in Ghana is grown exclusively in peasant farms. This also holds true for Salvador (Bahia) in Brazil but to a lesser degree. (see Table 40).

Table 40 also shows the wide spread of farm sizes in coffee growing in São Paulo in Brazil. In Africa coffee is also found in the widest variety of sizes of farm. In Angola, Kenya, Tanzania, the Congo region and parts of West Africa the plantation form of production prevails. On the other hand, Uganda and Ethiopia, which are counted among large exporters, have developed peasant coffee production. Small scale coffee

Table 39    Processing tea, coffee and cocoa crops.

| Tea | Coffee | Cocoa |
| --- | --- | --- |
| Pre-sorting; drying to 50–60%; rolling the leaves (mainly by machine); fermenting for 3 hours at high humidity at 23–25°C; drying at 80–100°C to 5–6%; sorting; packing | Separating the fruit and kernel of the berry in a running water pulper; fermentation; washing; natural or artificial drying; shelling and polishing by machine; sorting; shipping; roasting is carried out in the consuming country (There is also a lengthier dry processing method which improves quality) | Fermentation in bins for 3 to 4 days; drying to 3–4%; shelling by machines; removing sprouts from beans; grinding the cocoa; removing fat by pressing to separate cocoa butter and powder |

Source: Schütt (1972) pp. 166, 172, 179.

processing (dry processing) makes possible small scale production in peasant gardens. The less uniform quality of large quantities for export then tends to push down the price. The solution must be found in co-operatively organized large scale processing. This has been very successfully organized in Jamaica (Piekenbrock 1958).

Tea tends more to be a plantation crop as the increasing quality requirements on the world market can no longer be satisfied by hand preparation but require the facilities of large tea factories. Tea for developing countries does indeed continue to be grown on peasant holdings on quite a large scale (i.e. in Pakistan) but tea for industrial countries is becoming more and more a plantation crop (India, Sri Lanka, Indonesia). As Table 40 shows some 49% of the area under tea in Sri Lanka is in plantations larger than 200 ha and only 15% of tea is grown in small holdings up to 4 ha. In India 74.2% of the area under tea is in plantations above 500 acres (202 ha) in size.

## Labour economy at farm level

The tendency for tea to be grown in plantations does not arise only from the economic pressures to meet the demands of large scale factories within the smallest possible radius of production in countries where road and transport conditions are poor. It also arises because tea offers favourable possibilities for division of labour which can be exploited by using hired labour and made cheaper by mechanization. In the subtropics the harvest period lasts between five and ten months, and in the tropics as much as 12 months. As harvest labour accounts for 44% to 50% of total labour costs and as weeding which can be shifted to a convenient time accounts for another 15% to 22%, it is possible to even out the employment of the labour force. This is par-

Table 40    Enterprise size and labour economics for tea, coffee and cocoa plantations.

| Characteristic | Unit | Tropical Highlands | Tropical Lowlands | | |
|---|---|---|---|---|---|
| | | Tea | Coffee | | Cocoa |
| | | | arabica | robusta | |
| *A. area planted* | | Sri Lanka | São Paulo | | Salvador (Bahia) |
| ≤10 ha | | 15 (<4 ha) | 14 | | 23 |
| 10–100 ha | % cultivated | 36 (5–200 ha) | c60 | | 68 |
| >100 ha | area | 49 (>200 ha) | c26 | | 9 |
| suitable for | | plantations | plantations and peasant farms | | peasant farms |
| *B. labour* | | | | | |
| labour requirements | manhours/ha | 3 200–5 600 | 1 600–2 600 | 600–800 | 300–1 000 |
| harvest labour | % | 44–50 | c35 | 38–40 | c60 |
| length of harvest | months | ≤12 | 4 + 8[1] | 6 | 5 + 5[2] |
| labour costs | % total costs | c66 | c50 | c66 | c30 |

————————————— falling intensity of labour ————————————→

————————————rising labour productivity ———————————→

[1] 4 months main harvest and 8 months subsidiary harvest;
[2] 5 months each of main harvest and subsidiary harvest.
Sources: Franke (1967); Kullak-Ublick (1965); Metzner (1968); Ruthenberg (1971).

ticularly important because, as Table 40 shows, labour accounts for around two thirds of production costs. Generally tea is much the most labour intensive of this group of crops and it is therefore particularly suitable for growing in overpopulated developing countries like India or Ceylon.

Arabica coffee is less labour intensive. The harvest covers four months of main harvest and eight months of subsidiary harvest and absorbs a good third of the total labour requirement. As fertilizers and plant protection measures are already more important for quality coffee production, labour costs fall to around 50% of total costs.

For robusta coffee the share of harvest labour in total labour requirements rises, as does labour's share of total production costs. This is because land cultivation, fertilizers and plant protection are less important and because the very long economic life of the crop makes it easier to avoid the heavy cost of replanting. The total labour requirements for this relatively extensive type of coffee crop is only 30% to 40% of that for arabica coffee.

Cocoa is, or at least can be, an even more labour extensive crop. In Ghana the main and subsidiary harvests, each lasting about five months account for 60% of total labour requirements, which in turn account for 30% of total production costs.

Generally the crops in Table 40 are arranged from left to right in order of falling labour intensity and rising labour productivity.

Table 41   Optimal inputs of labour and fertilizers.

A. Labour intensity in growing cocoa in Nigeria[1]

| Level of Intensity | Input manhours/ha | Yield kg/ha | Labour costs DM/ha | Returns DM/ha | Returns less labour costs DM/ha |
|---|---|---|---|---|---|
| I | 172 | 226 | 115 | 490 | 375 |
| II | 301 | 309 | 202 | 670 | 468 |
| III | 511 | 346 | 343 | 750 | 407 |
| IV | 1 107 | 426 | 745 | 928 | 183 |
| V | 2 199 | 583 | 1 480 | 1 268 | −212 |

B. Inputs of nitrogen for growing coffee in Tanzania[2]

| Level of Intensity | Input kg/ha N | Yield kg/ha coffee | Production costs DM/ha | Returns DM/ha | Profit DM/ha |
|---|---|---|---|---|---|
| I | 20 | 500 | 1 214 | 1 375 | 161 |
| II | 70 | 840 | 1 875 | 2 310 | 453 |
| III | 100 | 960 | 2 100 | 2 640 | 540 |
| IV | 225 | 1 250 | 2 750 | 3 437 | 687 |
| V | 300 | 1 300 | 2 926 | 3 575 | 649 |

Basic data from: (1) Galleti et al (1956).   (2) Lentze (1969).

## Intensive or extensive production?

In spite of its relatively good labour productivity the growing of cocoa has been badly hit by wage levels in a country like Nigeria which has already got quite high wages for a developing country. Table 41 ranks labour input per ha with the corresponding yields. It demonstrates the effect of the law of diminishing returns and shows how the optimum intensity of production can be calculated by substituting cash values for physical inputs and outputs of labour and cocoa. Returns less labour costs in DM/ha reach their maximum at 301 manhours/ha and fall rapidly when the input of labour is more than 511 manhours.

The lower section of Table 41 shows the optimum input of nitrogen for growing coffee in Tanzania. This also demonstrates the law of diminishing returns, so that the highest profit in DM/ha occurs with an input of 225 kg/ha of pure N while both higher and lower inputs of N reduce profit. The optimal intensity of fertilizer application depends on the production function as well as on prices and costs. The optimal level of fertilizers in Tanzania certainly lies well above the world average. In the state of São Paulo the total application of all fertilizers in the largest plantations with over 512 000 coffee bushes lies at 152 kg/ha. The mean for all size classes is only 61 kg/ha. (Franke et al 1967 p.43). In the course of development, however, it is likely that the price of coffee will rise and the import price of mineral fertilizers will ease. At the very least the difference between the farm price of fertilizers and the price as it is landed in the port will fall as the infrastructure of roads and transport facilities is built up. The improvement of the price/cost relationship at the farm level will shift the margin of intensity of fertilizer application upwards and thus improve both production and incomes.

Finally a particularly pressing question of special intensity is that of the development of plant protection. Pests, diseases and weeds account for losses of 32.2% of the worlds potential tea yields, 44.4% of coffee and 45.9% of cocoa. The high cost of capital goods in developing countries and the difficulties in applying plant protection materials leave much to be desired in this field. However the prices of plant protection materials will certainly fall in the course of development and their method of application must be mastered by co-operative and supra-farm types of activity. Even now plant protection materials have the advantage for developing countries of low transport costs.

## Land and labour productivity

Finally Table 42 presents a productivity comparison that can only indicate fairly broad orders of magnitude because of the wide range of variability in all the data used in the estimates. As the amounts of farm inputs are largely unknown it is not possible to determine farm income and thus net productivity. The calculated and adjusted gross productivity is however a more informative indicator in developing countries than in developed countries as such inputs are so far extremely small.

Land productivity is a more important economic and competition indicator for thickly populated developing countries with rural underemployment and small farms. On this basis, as Table 42 shows, arabica coffee in Tanzania is far in the lead, followed by tea in Sri Lanka, robusta coffee in Brazil and lastly by cocoa in Nigeria. Thus tea is well suited not only to ecological conditions but also to economic conditions in India, Sri Lanka and China, while arabica coffee is similarly suited to Uganda and India. On the other hand thinly settled countries like Brazil, Indonesia and Zaire can accept a lower land productivity and produce robusta coffee and cocoa. These crops enable them to achieve high labour productivity which is more important in their circumstances than land productivity.

The model calculations demonstrate the well known antithesis between labour productivity and land productivity. One has to choose whether to put a high value on land or on labour and the decision must in principle be in favour of whichever is the more scarce and more expensive factor of production. As regards the four crops for which calculations have been carried out we can broadly say that the two upland crops have high land productivity but low labour productivity and are therefore suitable for overpopulated developing countries with low wage levels and small farms. The two lowland crops, on the contrary, combine a lesser land productivity

Table 42    Productivity in tea, coffee and cocoa plantations (model calculations).

| Indicator | Unit | Tropical uplands | | Tropical lowlands | |
| | | Tea | Coffee | Coffee | Cocoa |
| | | | (Arabica) | (Robusta) | |
| | | Sri Lanka | Tanzania | Brazil | Nigeria |
|---|---|---|---|---|---|
| A. Location | | | | | |
| Size of farm | ha | 200 | 80 | 200 | 3 |
| Height | m/OD | 1 300 | 1 800 | 400 | 200 |
| Precipitation | mm | 1 400 | 1 200 | 2 000 | 1 800 |
| B. Cultivation Cycle | years | 6 | 6 | 6 | 8 |
| Development Period | years | 50 | 24 | 100 | 40 |
| C. Expenditure | | | | | |
| Initial investment | DM/ha | 8 000 | 2 200 | 800 | 3 700 |
| Depreciation | DM/ha/year | 160 | 92 | 8 | 93 |
| Labour | manhours/ha/year | 3 200 | 1 900 | 740 | 500 |
| D. Returns | | | | | |
| Yields | q/ha | 15.6 | 8.4[1] | 8.5 | 6.0 |
| Producer price | DM/kg | 1.14 | 2.75 | 1.71 | 2.17 |
| Gross returns | DM/ha | 1 700 | 2 310 | 1 450 | 1 300 |
| E. Productivity | | | | | |
| Gross land productivity | DM/ha[2] | 1 610 | 2 218 | 1 442 | 1 207 |
| Gross labour productivity | DM/manhour[2] | 0.51 | 1.7 | 1.95 | 2.42 |
| Agricultural wage rate (for comparison) | DM/manhour | 0.23 | 0.49 | 0.52 | 0.67 |

1) With 70 kg/ha nitrogen.
2) Returns per ha or per manhour less depreciation.
Sources: FAO (1971); (1976); Irvine (1969); Kullak-Ublick (1965); Masefield (1948); Ruthenberg (1967 1971).

with a good labour productivity and are therefore destined for more thinly settled developing countries with higher wages and larger farms.

## Bananas and Citrus Fruits

### Geographical distribution of production

The range of citrus and other tropical fruits grown in different parts of the world is as a whole very varied. The range within individual Mediterranean and tropical regions is, however, much less, as ecological and to some extent economic production conditions for most citrus and other tropical fruits are fairly sharply defined.

Suitable conditions for growing bananas, mangoes and pineapples are found in the hot and humid equatorial zone; papayas and passion fruit prefer higher altitudes in tropical regions; dates predominate in the hot and dry northern tropical zone and in the subtropics conditions are favourable for all kinds of citrus fruits, fig-trees and avocados. If citrus fruits are grown in subtropical areas with changing humidity they need irrigation. The choice of citrus or other tropical fruit crops again depends also on economic criteria, as productivity conditions differ from one type of fruit to another. Thus, bananas which grow in bunches, provide by far the greatest food value per ha. They are, therefore, the best subsistence fruit crop in densely populated developing countries. On the other hand, they are not suitable for cultivation in industrial countries because of their extremely low labour productivity. Pineapples and citrus fruits are at the opposite extreme, their combination of high labour costs and high returns to land makes them ideal for industrial countries.

Table 43   Per caput consumption of citrus fruits and bananas.

| Country or Region | Citrus[1] | | | Country or Region | Bananas | | |
|---|---|---|---|---|---|---|---|
| | 1949/53 | 1961/63 | | | 1957/59 | 1964/66 | |
| | kg/year (=100) | Kg/year | index | | kg/year (=100) | kg/year | index |
| A. Producers | | | | | | | |
| Mediterranean & South Africa | 12 | 14 | 117 | East Asia & Pacific Is. | – | 7.6 | – |
| USA & Canada | 40 | 32 | 80 | Argentina[2] | 10.8 | 12.5 | 116 |
| Latin America | 24 | 22 | 92 | Portugal | 4.0 | 5.9 | 147 |
| B. Importers | | | | | | | |
| USSR & E. Europe | 28 | 12 | 43 | USA | 9.3 | 9.4 | 101 |
| UK | 11 | 13 | 118 | USSR | 0.02 | 0.09 | 450 |
| France | 11 | 16 | 146 | EEC | 5.3 | 7.7 | 145 |
| Germany FR | 6 | 18 | 300 | N. Africa & W. Asia | 1.0 | 1.4 | 140 |
| Scandinavia | 21 | 33 | 157 | South Africa | 6.2 | 5.7 | 92 |
| World Average | 6.3 | 6.6 | 105 | World Average[1] | 5.7 | 7.0 | 133 |

1) Approx. values calculated by the author.
2) Production plus imports.
Sources: Wolf (1965); Mackie & Falek; FAO (1973); UN (1974).

**The demand for bananas**

The following sections compare differences in per caput consumption over space and time for the two most important types of tropical and subtropical fruits, bananas and citrus fruits. The data for both fruits in Table 43 show no clear income elasticity of consumption.

In 1964/66 per caput consumption of bananas was higher in Argentina than in the United States while in East Asia it was almost as high as in the EEC. The reason for this is that bananas are a basic food in developing countries in humid tropical regions, but are only one among many varieties of fruit in northern industrial countries. In addition the high transport costs of exporting bananas lead to a considerable price differential between industrial and producer countries. Mackie (1971) points out that this particular extra cost has three consequences. Between 1948/52 and 1965 the proportion of world banana production which was exported rose only from 18% to 22%. Between 1964 and 1966, of the main banana producers Brazil exported only 4.8%, India 0.3%, Ecuador 40.2%, Cameroon 70.5%, Costa Rica 66.7% and the Honduras 58.4%. Per caput consumption is highest in less developed countries and can amount to between 25 and 45 kg per annum. In high-income importing countries consumption is much lower (9 to 10 kg per head per annum) and income elastic. Middle income level countries have the lowest per caput consumption of only 5 to 8 kg.

In practically all countries, the price elasticity of demand for bananas is higher than the income elasticity. Where income elasticity does have an effect, it becomes

Table 44   Income and price elasticity of demand for bananas and estimated growth of consumption from 1964/66 to 1980 in selected countries.

| Indicator | Italy | Netherlands | GFR | Sweden | USA |
|---|---|---|---|---|---|
| Income elasticity | 0.7 | 0.5 | 0.1 | 0.4 | 0.1[1] |
| Price elasticity | −1.1 | −0.7 | −1.1 | −0.7 | −0.1[1] |
| Population, × 10⁶ 1964/66 | 51.5 | 12.3 | 58.9 | 7.7 | 194.5 |
| 1980 | 56.8 | 14.2 | 62.3 | 8.7 | 241.1 |
| Per caput income, $[2] | | | | | |
| 1964/66 | 538 | 734 | 914 | 1132 | 2044 |
| 1980 | 983 | 1230 | 1554 | 1764 | 3048 |
| Per caput consumption, kg. | | | | | |
| 1964/66 | 5.2 | 6.8 | 9.5 | 7.8 | 7.9 |
| 1980 | 7.6 | 8.3 | 8.5 | 8.2 | 8.5 |
| Annual % increase 1964/80 | 2.5 | 1.3 | −0.5 | 0.3 | 0.5 |

1) North America as a whole. 2) Consumers' Spending Rate.
Source: Mackie & Falck (1971) pp. 69 and 71.

less as consumer income grows higher. As Table 44 shows, countries which start with relatively low incomes and have an absolutely smaller income growth such as Italy and the Netherlands are likely to show a much greater rise in per caput consumption of bananas than countries with higher initial incomes and greater absolute income growth such as Sweden and the USA.

## The demand for citrus fruits

The consumption of citrus fruit is subject to different laws. The price difference between producer and consumer countries is much less than for bananas. This is because citrus are technically easier to transport, the distances from producer to consumer countries is shorter and processing into concentrated juice and other preserved forms makes them still easier to transport.

As price varies less from one country to another income differences between countries have a much greater influence on consumption than in the case of bananas. In the German Federal Republic and in France the income elasticity of demand for oranges and tangerines amounted to 1.1 and 0.8 while the price elasticities of demand were −0.6 and −0.5 in 1963/64 (Wolf 1965). The ranking of elasticities is thus opposite to that of bananas.

The striking feature of Table 43 as regards citrus fruits is that there is a tendency for per caput consumption to fall after it reaches a certain level. In the Eastern European countries this may have been due to shortage of foreign exchange. In the case of the USA and even of Latin America this is clearly because consumption of citrus which is the dominant fresh fruit is reduced as consumers in the higher income bracket shift their consumption to a wider variety of competing fruits. In the same way richer consumers in the GFR tend to eat citrus fruits and bananas instead of apples. Per caput consumption of citrus fruits would show an even stronger decline in the highest income groups if it was not possible to process them into a number of different products the consumption of which is still tending to expand (see Table 45).

Table 45    Per caput consumption of citrus fruit.

| Use | Oranges | | | Grapefruit | | |
| --- | --- | --- | --- | --- | --- | --- |
| | 1945/46 | 1960/61 | 1969/70 | 1945/46 | 1960/61 | 1969/70 |
| Fresh fruit | 17.1 | 7.3 | 8.6 | 6.3 | 4.2 | 4.6 |
| Preserved fruits | 0 | 0 | | 0.5 | 1.0 | |
| Concentrated juice | | | 29.8 | | | 5.9 |
| Juice | 7.4 | 16.9 | | 7.9 | 2.1 | |
| Total Consumption | 24.5 | 24.2 | 38.4 | 14.7 | 7.3 | 10.5 |

Sources: Citrus Fruits (1972); Reuther et al (1967).

## Seasonal supply and demand

The fact that harvest times in the southern hemisphere are six months later than in the northern tends to increase the consumption of all citrus and other subtropical fruits. It ensures that world market supplies are spread over a longer period, sometimes right throughout the year. Table 46 shows this clearly. Out of season production of oranges in Brazil, South Africa and Australia completely closes the gap in northern hemisphere supplies between July and October. Thus throughout the year citrus fruits are being harvested somewhere in the world and supplied to consumers.

This regional market adjustment is not needed for bananas because the main producer countries (Brazil, India, Ecuador, Cameroon) lie close to the equator in humid tropical zones with no distinct seasonal changes so that bananas can be harvested all through the year.

While pineapples and citrus are typical export crops and mangoes, dates and papayas are typical subsistence fruits and bananas are suitable for both purposes. The demand for bananas is more price elastic, that for citrus fruits more income

Table 46   The seasonal supply of oranges from selected producer countries.

| Month | California | Spain | Israel | Italy | Brazil | South Africa | Australia |
|---|---|---|---|---|---|---|---|
| November | X | X | X | | | X | X |
| December | X | X | X | | | | |
| January | X | X | X | X | | | |
| February | X | X | X | X | | | |
| March | X | X | X | X | | | |
| April | X | X | X | X | X | | |
| May | X | X | | X | X | X | |
| June | X | X | | X | X | X | X |
| July | X | X | | | X | X | X |
| August | X | | | | X | X | X |
| September | | | | | X | X | X |
| October | | X | | | X | X | X |

Source: Delfs-Fritz (1970), p. 136.

elastic. In a developing world economy it is, therefore, much easier for citrus growers to achieve an increase in wage and income levels than it is for banana growers.

If account is also taken of the specific productivities of both crops, there seems likely to be much more favourable long term prospects for the development of world citrus production than for world banana production. In the light of the problems of maintaining soil fertility in the humid tropics, this is a situation which is all the more to be regretted.

# Bibliography

Anon (1963) [Wood, raw rubber and copra, North Borneo's life-line. Energetic efforts to expand the bases of the economy.] Holz, Rohgummi und Kopra, Nordborneos Lebensader. Energische Bemühungen zur Erwieterung der Wirtschaftsgrundlagen. *Asien-Wirtschaft, Atlantic Union of Economic Geographers* 2 (2) 55 et seq.

Albrecht, H. (1964) [The extension service's socio-psychological interpretation of the symbiosis between cattle farming and coconut palm cultivation.] Sozial-psychologische Interpretation des Beratungsdienstes über Symbiose zwischen Rindviehhaltung und Kokospalmenanbau. *Zeitschrift für Ausländische Landwirtschaft* 3, 70–71.

Allen, I.L. (1966) The market for kapok fibre and seed. *Report, Tropical Products Institute* No. G 27, 23pp.

Allison, H.W.S.; Smith, R.W. (1964) Economics of cacao establishment on clear field land. *World Crops* 16 (3), 31–36.

Andreae, B. (1974) [Sugar beet and sugar cane in competition. Development problems of the sugar industry in Khuzestan, Iran.] Zuckerrübe und Zuckerrohr im Wettbewerb. Entwicklungsprobleme der Zuckerwirtschaft in Khuzestan/Iran. *Zeitschrift für die Zuckerindustrie* 24, 359–368.

Andreae, B. (1975) Natural fibre production in world agriculture. Economic scope within ecological limits. *Plant Research and Development* 1, 70–90.

Andreae, B. (1976) The influence of prices and income on consumption growth rates: elasticity of demand for bananas and citrus fruits. *Economics* 14, 123–128.

Andreae, B. (1977) [Agricultural geography.] Agrargeographie. Berlin; New York, USA; Walter de Gruyter 332pp. (English language edition forthcoming, New York, Walter de Gruyter 1980).

Attems, M.; Ruthenberg, H. (1969) [Systems and characteristics of mixed cropping in the tropics.] Systematik und Merkmale der Mischkultur in den Tropen. *Zeitschrift für Ausländische Landwirtschaft* 8, 2–8.

Aubert, H.J. (1974) [Development of the plantation sector in Sri Lanka within the framework of the agricultural sector after achieving independence.] Die Entwicklung der Plantagenwirtschaft Sri Lanka im Rahmen der Agrarwirtschaft nach der Erlangung der Unabhängigkeit. Dissertation, Universität Köln, German Federal Republic 223pp.

Blanckenburg, P. von (1965) [African peasant farming on the way to modern agriculture.] Afrikanische Bauernwirtschaften auf dem Weg in eine moderne Landwirtschaft. *Zeitschrift für Ausländische Landwirtschaft* Spec. No.3, 111pp.

Blücher, N. von (1956) [Tea. Cultivation and fertilizer use.] Tee. Anbau und Düngung. *Schriftenreihe über Tropische und Subtropische Kulturpflanzen, Ruhr-Stickstoff AG* 84pp.

Borchert, G. (1972) [Economic regions of the Ivory Coast.] Die Wirtschaftsräume der Elfenbeinküste. *Hamburger Beiträge zur Afrika-Kunde* No.13, 174pp.

Brown, D. (1971) Agricultural development in India's districts. Cambridge, Massachusetts, USA; Harvard University Press 169pp.

Butterwick, M. (1964) Prospects of Indian tea exports. *International Journal of Agrarian Affairs* 4 (4), 205–249.

Caesar, K. (1969) [Possibilities of introducing new crops into the tropics and sub-tropics.] Möglichkeiten der Einführung neuer Kulturpflanzen in den Tropen und Subtropen. *Zeitschrift für Ausländische Landwirtschaft* 8, 260–272.

Commerzbank, Berlin (1967) [Problems on the international cocoa market. Worldwide efforts to achieve price stability.] Probleme des internationalen Kakaomarktes. Weltweite Bemühungen um Preisstabilität. *Aussenhandelsblätter* 19.

Commonwealth Economic Committee, Intelligence Branch (1960 and 1963) Plantation crops. A review of production, trade, consumption and prices relating to sugar, tea, coffee, cocoa, spices, tobacco, and rubber.

Cookman, G.P. (1970) Bananas in Ecuador. In: Bunting, A.H. (*Ed*) Change in agriculture. London, UK; Gerald Duckworth 167–174.

Delfs-Fritz, W. (1970) Citrus. Cultivation and fertilizer use. *Series of Monographs on Tropical and Subtropical Crops* 230pp.

Dumont, R. (1957) Types of rural economy: Studies in world agriculture. London, UK; Methuen & Co. 555pp.

FAO (1960) Tea—trends and prospects. *Commodity Bulletin Series* No.30.

FAO (1966) The longer-term outlook for cocoa production and consumption. *Monthly Bulletin of Agricultural Economics and Statistics* 15.

FAO (1970, 1972 and 1977) Production Yearbook 23 (1969); 25 (1971); and 30 (1976).

FAO (1969) Tanzanian sisal. *Monthly Bulletin of Agricultural Economics and Statistics* 18 (5), 1–8.

FAO (1974) Developments and problems in the world banana market—possible international approaches. *Monthly Bulletin of Agricultural Economics and Statistics* 23 (9) p.21 et seq.

Franke, G. (et al) (1967) [Useful crops of the tropics and sub-tropics.] Nutzpflanzen der Tropen und Subtropen, Vols.I and II. Leipzig, German Democratic Republic; S.Hirzel Verlag 324 pp.; 421pp.

Galletti, R.; Baiduin, K.D.; Dina, J.O. (1956) Nigerian cacao farmers. London, UK; Oxford University Press 744pp.

Geer, T. (1970) The international market for natural rubber. *Zeitschrift für Ausländische Landwirtschaft* 9, 211–244.

Gerling, W. (1954) [Plantation.] Die Plantage. Ed.2, revised. Würzburg, German Federal Republic; Stahel'sche Universitätsbuchhandlung 47pp.

Gnielinski, S. von (1968) [Sugar cane cultivation in Liberia and its economic significance.] Zuckerrohranbau in Liberia und seine wirtschaftliche Bedeutung. *Zeitschrift für Ausländische Landwirtschaft* 7, 276–291.

Harler, C.R. (1964) The culture and marketing of tea, Vol.XI, Ed.3. London, UK; Oxford University Press 262pp.

Heinemann, C. (1953) [India rubber. Cultivation and fertilization.] Kautschuk. Anbau und Düngung. *Schriftenreihe über Tropische und Subtropische Kulturpflanzen, Ruhr-Stickstoff AG* 113pp.

Hilkenbäumer, F.; Schmitz-Hübsch, E. (1971) [Cost accounting in commercial fruit growing.] Kalkulation im Erwerbsobstbau. *Gärtnerische Berufspraxis* No.31, Ed.2, Fully Revised 130pp.

Irvine, F.R. (1959) West African crops. Oxford, UK; Oxford University Press 270pp.

Jacoby, T. (1954) [The oil palm. Cultivation and fertilization.] Die Ölpalme. Anbau und Düngung. *Schriftenreihe über Tropische und Subtropische Kulturpflanzen, Ruhr-Stickstoff AG* 49pp.

Jensch, G. (1970) [Global climate.] Klima-Globus. Berlin; Niepert KG 50pp.

Jepson, W.F. (1956) Methods of plant protection for peasants in the tropics. *Outlook on Agriculture* 1 (2), 59–63.

Juelich, V. (1975) [Agricultural colonization of the rain forest of the middle Rio Huallaga (Peru).] Die Agrarkolonisation im Regenwald des mittleren Rio Huallaga (Peru). *Marburger Geographische Schriften* No.63, 236pp.

Koelle, A. (1967) [Cultivation, harvesting and packaging of Chiquita bananas in Honduras.] Anbau, Ernte und Verpackung der Chiquita-Banane in Honduras. *Tropenlandwirt* 68, 103–108.

Könnecke, G. (1967) [Crop rotation.] Fruchtfolgen. Berlin; VEB Deutscher Landwirtschaftsverlag 335pp.

Kullak-Ublick, H. (1965) [The tea sector in Ceylon.] Die Teewirtschaft in Ceylon. *Zeitschrift für Ausländische Landwirtschaft* 4, 33–45.

Lamade, W. (1968) Marketing boards in Tanzania. *Zeitschrift für Ausländische Landwirtschaft* 7, 334–348.

Lentze, W. (1969) The combination of factors of production in developing countries.] Gedanken zur Produktionsfaktorenkombination in Entwicklungsländern. *Tropenlandwirt* 70, 149–156.

Mackie, A.B.; Falck, J.F. (1971) World demand prospects for bananas in 1980. *Foreign Agricultural Economic Report, Economic Research Service, US Department of Agriculture* No.69, 94pp.

Masefield, G.B. (1948) The life of perennial crops. *East African Agricultural Journal of Kenya.*

Massow, H. von (1963) [Coffee in Tanganyika. Situation and development possibilities.] Kaffee in Tanganyika. Situation und Entwicklungsmöglichkeiten. Hamburg, German Federal Republic; Afrika-Verein, Technisch-Wirtschaftlicher Dienst 30pp.

Massow, H. von (1965) [Peasant tea in Kenya.] Bauern-Tee in Kenya. *Afrika Heute, Deutsche Afrika-Gesellschaft* 21, 291–292.

Maydell, H.J.; Erichsen, H. (1968) [Incidence and use of economically important palms.] Vorkommen und Nutzung wirtschaftlich wichtiger Palmen. *Mitteilungen der Bundesforschungsanstalt für Forst- und Holzwirtschaft* No.69, 142pp.

Melville, A.R. (1970) The development of coffee production by African farmers in Kenya. In: Bunting, A.H. (*Ed*) Change in agriculture. London, UK; Gerald Duckworth 119–130.

Metzdorf, H.J. (1964) [The world market for fats.] Der Weltmarkt für Fette. *Agrarwirtschaft* 13, 379–380.

Metzner, U. (1968) [The cacao zone of Bahia.] Die Kakaozone der Bahia. *Zeitschrift für Ausländische Landwirtschaft* 7, 196–207.

Mylord, E. (1953) [Cacao. Cultivation and fertilizer use.] Kakao. Anbau und Düngung. *Schriftenreihe über Tropische und Subtropische Kulturpflanzen, Ruhr-Stickstoff AG* 109pp.

Niederstucke, H. (1970) [Forms of land use in tropical highlands.] Bodennutzungsformen in tropischen Höhenlagen. *Landwirt im Ausland* 4, 74–76.

Nitz, H.J. (1975) [The economic region and economic structure.] Wirtschaftsraum und Wirtschaftsformation. In: Der Wirtschaftsraum. *Geographische Zeitschrift Beihefte* 41, 42–58.

Ocker, W. (1965) [Modern methods of tea processing.] Neuzeitliche Methoden der Teeaufbereitung. *Tropenlandwirt* 66, 68–75.

Padwick, G.W. (1970) Cocoa spraying in Ghana and Nigeria. In: Bunting, A.H. (*Ed*) Change in agriculture. London, UK; Gerald Duckworth 615–624.

Piekenbrock, P. (1958) [Vegetation and crop cultivation in the tropics.] Vegetation und Pflanzenbau in den Tropen. *Schriftenreihe der Deutschen Afrika-Gesellschaft* No.7, 36pp.

Poerck, R. de (1965) [International contribution to improvement of the palm oil industry.] Internationaler Beitrag zur Verbesserung der Ölpalmen-Wirtschaft. *Zeitschrift für Ausländische Landwirtschaft* 4, 303–304.

Pössinger, H. (1967) [Sisal in East Africa. Investigations of productivity and profitability in peasant farming.] Sisal in Ostafrika. Untersuchungen zur Produktivität und Rentabilität in der bäuerlichen Wirtschaft. *Afrika-Studien, IFO-Institut für Wirtschaftsforschung* No.13, 172pp.

Reuther, W.; Webber, H.J. et al (*Eds*) (1967) The citrus industry, Vol.V. USA; University of California.

Ruhr-Stickstoff, A.G. (1953–1972) *Schriftenreihe über Tropische und Subtropische Kulturpflanzen.*

Rump, K. (1969) [Problems of monoculture, illustrated by examples from Ecuador's banana sector.] Probleme der Monokultur, dargestellt am Beispiel der Bananenwirtschaft Ecuadors. *Arbeitsberichte des Ibero-Amerika-Instituts für Wirtschaftsforschung an der Universität Göttingen* No.4, 174pp.

Ruthenberg, H. (1963) [Starting points for and hinderances to further agricultural development in Madagascar. Notes from a journey.] Ansätze und Hindernisse der Weiteren landwirtschaftlichen Entwicklung in Madagaskar. Notizen einer Reise. *Zeitschrift für Ausländische Landwirtschaft* 2, 18–59.

Ruthenberg, H. (1967) [Ways of organizing land use and livestock farming in the tropics and sub-tropics, demonstrated by selected examples.] Organisationsformen der Bodennutzung und Viehhaltung in den Tropen und Subtropen. In: Von Blanckenburg, P.; Cremer, H.D. (*Eds*) Handbuch der Landwirtschaft und Ernährung in den Entwicklungsländern Vol.I. Stuttgart, German Federal Republic; Eugen Ulmer 122–208.

Ruthenberg, H. (1976) Farming systems in the tropics. 2nd edn. Oxford, UK; Clarendon Press 366pp.

Schendel, U. (1970) [Report on a water management study and Congress trip to Australia, New Zealand, Hawaii and California from 10th Nov. to 22nd Dec., 1970.] Bericht über eine wasserwirtschaftliche Studien- und Kongressreise nach Australien, Neuseeland, Hawaii und Kalifornien vom 10. Nov. bis 22. Dez. 1970. Kiel, German Federal Republic; Manuscript 47pp.

Schönwälder, H. (1969) [The cocoa sector in West Africa.] Die Kakao-Wirtschaft in Westafrika. *Hamburger Beiträge zur Afrika-Kunde* No.9, 232pp.

Schütt, P. (1972) [World industrial crops.] Weltwirtschaftspflanzen. Hamburg, German Federal Republic; Paul Parey 228pp.

Schweinfurth, U. (1966) [The tea area in the highlands of the island of Ceylon as an example of landscape change.] Die Teelandschaft im Hochland der Insel Ceylon als Beispiel für den landschaftswandel. *Heidelberger Studien zur Kulturgeographie*, Festausgabe zum 65, Geburtstag von G.Pfeifer 15, 297–310.

Steinhausen, W. (1957) [The banana. Cultivation and fertilizer use.] Die banane. Anbau und Düngung. *Schriftenreihe über Tropische und Subtropische Kulturpflanzen, Ruhr-Stickstoff AG* 104pp.

Stern, R.M. (1965) The determinants of cocoa supply in West Africa. In: Stewart, I.G.; Ord, H.W. (*Eds*) African primary products and international trade. Edinburgh, UK; Edinburgh University Press 65–82.

Strenge, H. von (1954) [Coffee. Cultivation and fertilizer use.] Kaffee. Anbau und Düngung. *Schriftenreihe über Tropische und Subtropische Kulturpflanzen, Ruhr-Stickstoff AG* 111pp.

Stroebel, H. (1975) [Development opportunities for small farms in the Kericho District of Kenya.] Entwicklungsmöglichkeiten landwirtschaftlicher Kleinbetriebe im Kericho Distrikt, Kenia. Stuttgart-Hohenheim, German Federal Republic. (Manuscript).

Tanner, H.J. (1968) [Colombian coffee.] Der kolumbianische Kaffee. *Geographice Helvetica, Geographisch-Ethnographische Gesellschaft* 23, 180–186.

Taylor, J.A. (1967) World situation and outlook for bananas. In: Agricultural producers and their markets. Oxford, UK; Basil Blackwell 273–284.

UN, Statistical Office (1974) Statistical Yearbook, New York, USA 829pp.

USA, Department of Agriculture (1971) World demand prospects for bananas in 1980. *Foreign Agricultural Economic Report, Economic Research Service* No.69, 94pp.

USA, Department of Agriculture (1972) Citrus fruits. Revised estimates by States, 1964/65 to 1969/70. *Statistical Bulletin, Statistical Reporting Service* No.493, 13pp.

Walker, H. (1975) [The oil palm cultivation sector in Sumatra and West Africa.] Die Ökonomie des Ölpalmenbaues in Sumatra und Westafrika. Stuttgart-Hohenheim, German Federal Republic. (Manuscript).

Wilkens, P.J. (1974) [Changes in the plantation sector. Decolonization of one form of the economy.] Wandlungen der Plantagenwirtschaft. Die Entkolonisierung einer Wirtschaftsform. Dissertation, Universität Hamburg, German Federal Republic 173pp.

Wolf, J. (1965) The citrus economy and the feasibility of international market arrangements. *Monthly Bulletin of Agricultural Economics and Statistics* 14 (9) 1–15.

Wood, G.A.R. (1970) Contribution of manufacturers to cocoa production in West Africa. In: Bunting, A.H. (*Ed*) Change in agriculture. London, UK; Gerald Duckworth 293–298.

Wrigley, G. (1971) Tropical agriculture. The development of production. London, UK; Faber 376pp.

# Chapter 8

# Development of Farming Regions and Farm Enterprises during General Economic Growth

Almost all agricultural countries suffer from a considerable imbalance between the three production factors land, labour and capital. These disproportions in factor endowment make it difficult to achieve the best combination for production purposes and as a result their per caput income remains unsatisfactorally low and they have the characteristics of a developing country.

All developing countries suffer from a shortage of capital, and the overpopulated ones also suffer from a shortage of land. Nyasaland had too many people, Southern Rhodesia too much land and Northern Rhodesia was relatively well endowed with capital. The federation of Rhodesia and Nyasaland brought the factors of production more into balance and was therefore economically effective. When it collapsed politically, it left overpopulated Malawi and underpopulated Rhodesia with corresponding economic difficulties.

The terms "overpopulated" and "underpopulated" must, of course, be understood in a relative sense. They mean that given a particular set of natural resources and capital endowment, a more favourable combination of production factors, and hence higher income per caput, could be achieved with a lower or higher population density. West Germany was overpopulated before the last war and is today underpopulated, although the population density has increased considerably. Industrialization increased the GFR's production capacity so effectively that in 1970 some 1.58 million non-German workers were employed there as contract labour.

## Aspects of Economic Growth that are Significant for Agriculture

The task of every development policy is step by step to correct and eliminate this dis-equilibrium between factors of production. In overpopulated developing countries, attempts are made to improve the relationship between population and the basic food supply by encouraging birth control, although rather unsuccessfully, and by reclaiming and opening up new land and increasing yields on existing cultivated land.

**Industrialization**

Land reclamation and yield increasing measures in themselves increase the demand for capital and the need to make capital cheaper. Only in exceptional cases can land be reclaimed by the use of labour alone, and above a certain limit yield increasing measures require irrigation, fertilizers, plant protection etc. i.e. they need capital.

Industrialization helps to bring about capital accumulation and to reduce the cost of capital because industry is in a much better position than agriculture to take advantage of the specialization and concentration of production which increase productivity. Industry can also create new jobs and thus reduce rural overpopulation. This is a continuous process. Although industrial jobs initially create equilibrium between land and labour in rural areas, further industrialization reduces the cost of capital goods used in agriculture, and thus increases the capital intensity of agriculture and releases more labour. This new disequilibrium between land and labour in rural areas, must then be eliminated by further industrialization, and the cycle continues. Industrialization also helps developing countries by making them less dependent on imports of finished industrial products. This helps to relieve their food situation, either by reducing the need for agricultural exports to earn foreign exchange or by earning foreign exchange from industrial exports which can be used to buy agricultural imports.

Any degree of industrialization has a favourable effect upon market prices of farm goods and hence upon farm gate prices. Industrialization increases the purchasing power of the urban population, this improves the prices of agricultural products and particularly of livestock products. Industrialization encourages macro-economic specialization, and this is reflected in generally lower costs of production which affect farm machinery and equipment, fertilizers and other working stock of the farmer. Any industrialization creates incentives for intensification, lowers the cost of capital and thereby accelerates the development of more productive forms of agriculture (see Table 47).

## Development of infrastructure

The most urgent and important development measures in the agricultural sector seem to lie in the sphere of infrastructure policy. A wise infrastructure policy can speed up the evolution of farming systems by accelerating or initiating the shift in factor-cost relationships which is really the basis of overall economic development. This is possible even before any change in market prices takes place at all. The farmer, of course, is not concerned with market prices, but with farm gate prices. These differ considerably in the tropics, because of the undeveloped transport systems (cf. Fig.32).

The farmer has to meet transport costs from the farm to the market for his own products and the price he receives is therefore the market price less transport costs. On the other hand, he has to pay market price plus transport costs for industrially produced capital goods. Thus, the further a farm is from the market, the lower are the farm gate prices for its products, and the higher the prices of industrially produced inputs delivered to the farm. It must therefore be farmed more extensively and in particular, the employment of more capital is inhibited. Any development of transport lowers these costs, and improves the exchange relationships between farm products and industrial farm inputs. In this way every new railway, every new road, every fall in transport costs creates incentives to intensification and increased use of capital, and helps agriculture to develop in the desired direction.

## Off-farm inputs, farm sizes and other operational factors

When agricultural policy extends and improves the rural road network this has the same effects upon the change in the forms of farming as a general infrastructure policy. If the government or private corporations build oil mills, cane sugar and sisal factories, citrus fruit packing plants, etc., these establish new, closer marketing

Table 47   Agriculture in the economic growth of selected developing countries.

| | Sparsely settled developing countries | | | | | | Densely populated developing countries | | | | | |
| | very poor | | | less poor | | | very poor | | | less poor | | |
| | Egypt | Thailand | Indonesia | Brazil | Malaysia | Costa Rica | India | Ceylon | R. Korea | Philippines | Jamaica | Taiwan |
|---|---|---|---|---|---|---|---|---|---|---|---|---|
| Population density per km² (1966)[5] | 30 | 61 | 72 | 10 | 28 | 29 | 163 | 175 | 295 | 112 | 168 | 356 |
| Income per caput[2] (US$ in 1965)[5] | 110 | 113 | 82 | 232 | 272 | 382 | 92 | 137 | 93 | 237 | 453 | 200 |
| **Percentage aggregate growth rates[6]** | | | | | | | | | | | | |
| Agricultural production | 2.4 | 4.5 | 2.3 | 3.8 | 4.1 | 4.2 | 2.6 | 2.9 | 3.7 | 3.7 | 2.7 | 4.4 |
| Population | 2.5 | 3.3 | 2.2 | 3.0 | 3.1 | 3.8 | 2.2 | 2.4 | 2.8 | 3.3 | 1.8 | 3.2 |
| Agricultural production per caput | -0.1 | 1.2 | 0.1 | 0.8 | 1.0 | 0.4 | 0.4 | 0.5 | 0.9 | 0.4 | 0.9 | 1.2 |
| Real income per caput | 4.1 | 2.8 | -0.2 | 2.3 | 2.3 | 1.5 | 1.9 | 0.8 | 3.4 | 2.2 | 5.1 | 4.8 |
| Income elasticity of demand for farm products | 0.7 | 0.7 | 0.7 | 0.5 | 0.5 | 0.5 | 0.7 | 0.7 | 0.7 | 0.7 | 0.4 | 0.7 |
| Domestic demand for farm products | 5.4 | 5.3 | 2.1 | 4.2 | 4.3 | 4.5 | 3.5 | 3.0 | 5.1 | 4.8 | 3.8 | 6.6 |
| Deficit in farm production[3] | 3.0 | 0.8 | -0.2 | 0.4 | 0.2 | 0.3 | 0.9 | 0.1 | 1.4 | 1.0 | 1.1 | 2.2 |
| **Percentage growth rates in farming[6]** | | | | | | | | | | | | |
| Cultivated area | 0.6 | 2.4 | 2.2 | 4.1 | 1.6 | 2.3 | 1.3 | 1.1 | 1.5 | 2.7 | 2.8 | 0.7 |
| Yield per ha. | 2.3 | 2.1 | 2.1 | 0.2 | 2.1 | 1.4 | 1.5 | 2.1 | 2.8 | 1.4 | 0.6 | 3.1 |
| Farm population[4] | 2.2 | 2.9 | 2.9 | 1.3 | 3.1 | 2.3 | 2.1 | 1.2 | 0.3 | 2.0 | 1.3 | 2.4 |
| Farm production per member of the farm population | 0.2 | 1.6 | 1.6 | 2.5 | 1.0 | 1.9 | 0.5 | 1.7 | 3.4 | 1.7 | 1.4 | 2.1 |
| Area per member of the farm population | -1.2 | 0.6 | 0.8 | 3.1 | -1.5 | 2.1 | -0.8 | 0.3 | 1.2 | 0.7 | 1.6 | -1.7 |

1) Partly for a shorter period;
2) Gross social product at factor cost;
3) Growth in farm production deducted from growth in domestic demand for farm products;
4) Estimated from various sources, numbers employed in agriculture were recorded for Ceylon, Costa Rica, Indonesia, Malaysia, Thailand and Egypt;
5) Institut für Ausländische Landwirtschaft der TU Berlin (1969);
6) USDA. ERS (1970).

Fig. 32
Farm-gate price-cost relationships as distance from the market increases.

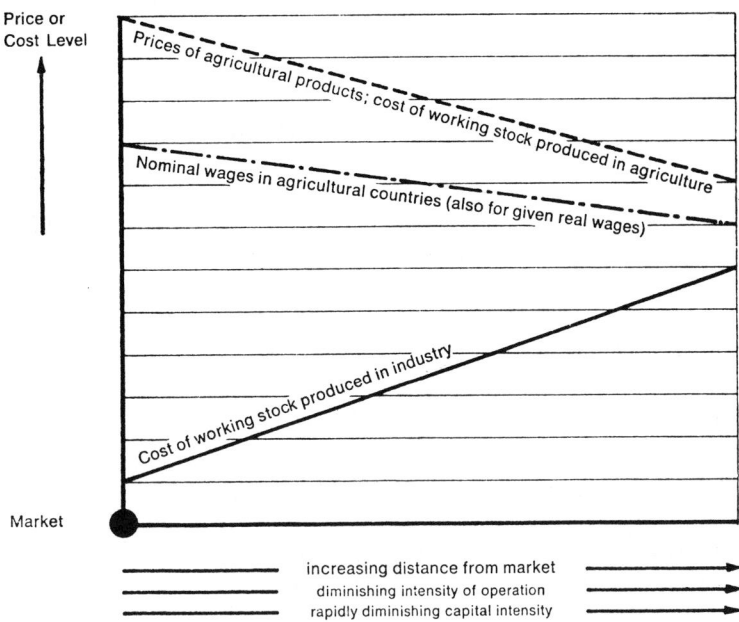

centres for the surrounding rural area and strongly influence the direction and intensity of production. If the government drills wells in unproductive areas of thornbush and semi-desert, land which previously could not even be used by the nomads as it lacked drinking water now becomes available to enlarge the feed-base for the herds, which are often too large. In other cases, wells can make pastures usable by sheep or cattle, which had previously only been accessible to camels because they were too remote from a watering place. If an organization can be created to transform an uncontrolled forest burning system into a planned one coordinated between all farmers, this produces a considerable improvement in agricultural structure. The Community Development Programme has been fruitful in this way.

Agricultural structure policy is usually understood to mean primarily land reform. Unfortunately the motives for this are often political rather than economic, and are therefore outside the scope of this book. Politically induced land reform usually takes place in the form of a revolution, and agriculture needs evolution rather than revolution. This also applies to farm size structure which must change to some extent in the course of overall economic development.

## International exchange of goods

The example of the former federation of Rhodesia and Nyasaland showed that political units which are too small make it difficult to develop the specialization which increases productivity. Large countries such as the USA, the USSR or Brazil are at an advantage here and can farm in a way more suited to the geographical location.

For this reason alone, other things being equal, small countries are more likely to be obliged to exchange goods with other countries than are large countries. This applies particularly to many tropical countries, which in some cases provide luxuries and semi-luxuries which developing countries consume only on a small scale and which industrial countries cannot produce (coffee, cocoa, pineapple, rubber). Thus Table 48 shows that Mexico exports the following percentages of her agricultural production: cocoa 27.2%, coffee 52.1%, raw sugar 24.8% and cotton 86.6%.

Agriculture then is part of the great international exchange of goods which also tends to increase its productivity continually. If the scope of specialization is ultimately extended considerably beyond national boundaries by a wise foreign trade policy, this will bring the time nearer when farmers in the humid tropics can return to a natural vegetation cover. They can then specialize for the world as a whole on the large scale cultivation of those prennial crops for which the humid tropics are a preferred location. This will make possible far more complete resource utilization and maintenance of soil fertility than hitherto.

Table 48    Production and export of tropical products in Latin America 1970.

| Product | Mexico | | | Brazil | | | Latin America as a whole | | |
|---|---|---|---|---|---|---|---|---|---|
| | Production | Export | Surplus | Production | Export | Surplus | Production | Export | Surplus |
| | t × 10³ | t × 10³ | % | t × 10³ | t × 10³ | % | t × 10³ | t × 10³ | % |
| Bananas | 1 136 | 5 | 0.4 | 9 600 | 200 | 20.4 | 20 042 | 5 096 | 25.4 |
| Cacao | 22 | 6 | 27.2 | 175 | 120 | 68.6 | 355 | 160 | 45.0 |
| Coffee | 192 | 100 | 52.1 | 585 | 1 026 | 175.0 | 1 890 | 2 010 | 106.0 |
| Raw sugar | 2 373 | 590 | 24.8 | 4 990 | 1 130 | 22.7 | 13 480 | 4 533 | 33.6 |
| Rice | 330 | 26 | 7.9 | 7 350 | 91 | 12.4 | 10 787 | 216 | 2.0 |
| Maize | 8 200 | −700 | – | 14 200 | 1 470 | 10.0 | 36 851 | 5 457 | 14.8 |
| Cotton | 341 | 295 | 86.6 | 580 | 345 | 59.5 | 1 493 | 872 | 58.5 |
| Tobacco (raw) | 63 | 8 | 12.7 | 196 | 51 | 26.0 | 456 | 107 | 23.4 |
| Beef and veal | 604 | 50 | 8.3 | 1 650 | 111 | 6.7 | 6 541 | 874 | 13.3 |

Source: USDA, ERS (1971).

## Agricultural price level and structure

Even this outline of overall economic growth processes enables us to draw substantial conclusions on the evolution and adaptation of tropical agriculture as these processes radically affect the level and structure of agricultural prices. Economic growth has the following effects:

1. Industrialization increases personal incomes of the mass of the population, this increases the demand for food and this, in turn, increases the general level of agricultural prices. Other circumstances being equal, this must increase the exchange value of agricultural products against wages and farm inputs. As industrialization simultaneously reduces the prices of industrially produced farm inputs and, because of competition for labour, pushes up wage levels, it produces the following changes in the organization of agricultural production:

    (a) The intensity of organization increases as intensive types of farming expand at the expense of extensive types and as fallow periods are reduced.

(b) Capital is substituted for labour. In sparsely populated countries the emphasis is on labour saving capital goods (machines for tillage, preservation of soil fertility, harvesting, transport, and processing equipment); in densely populated countries the emphasis is more on yield increasing inputs (irrigation, fertilizers, plant protection, veterinary hygiene, possibly feed concentrates).

(c) The ratio between the rise in wages and the fall in the cost of capital will determine whether the overall specific intensity is increased—i.e. whether the reduction of labour in an individual branch of farming is more than offset by the additional use of yield increasing inputs. In densely populated developing countries, the answer to this question will always be in the affirmative, but in sparsely populated countries this will not always be so.

2. After a certain stage of development is reached, growing prosperity and associated changes in consumption habits will push up the prices of animal products distinctly more than those of crop products. This promotes livestock husbandry in its existing locations and encourages its introduction in new locations. Savannah farms that have hitherto been without livestock are turned into mixed farms which integrate arable and stock farming, supporting each other with fodder crops and manure. This intimate symbiosis between arable and stock farming then appears which has helped Central European agriculture to reach such a high level of productivity.

3. Improvements in infrastructure and in farm sizes also lower costs. They improve the relationship between costs and output and hence also increase the intensity of agriculture. In addition the increase in the number of markets and in the density of the transport network have a levelling effect on agriculture. Previous regional differences due to distance from the market are eliminated as transport costs as a whole fall and the burden of transport costs tends to become equalized for all farms. The more this happens, however, and the weaker the influence of the external transport situation becomes, the more agriculture is able to adapt itself to the natural production conditions in any location.

## Factor Costs and Factor Combination

If the process of overall economic growth is conceived as a whole right up to its highest stage, then four stages of development can be roughly distinguished:
1. Agricultural countries;
2. Agricultural industrial countries;
3. Industrial agricultural countries;
4. Industrial countries.

During this development process factor costs and factor combination change differently in sparsely populated countries and densely populated countries. Chapter 2 has already summarized what has been said about the minimum cost combination in agriculture in the course of economic development and examples are shown in Figure 33. In agriculture three elementary production factors are distinguished, viz. land, labour and capital. Technically, they do not need to be combined in any particular proportions and are, indeed, interchangeable within quite wide limits. In the economic sense, however, there is only one minimum cost combination of these three production factors for every stage of overall economic development. This is based on

Fig. 33
Factor costs and combination in the course of economic development. (Source: Herlemann 1954, p.335 ff)

| Typical examples | Thinly-populated Countries | | | Trend of Development | Densely-populated Countries | | | Typical examples |
|---|---|---|---|---|---|---|---|---|
| | Land | Labour | Capital | ↓ | Land | Labour | Capital | |
| | *A. Costs Ratio* | | | | | | | |
| Zaire, Ethiopia | – – | + | + + | Agric. country | + + | – – | + + | UAR, Ghana, Thailand |
| Mexico, Zambia, Kenya | – | + | + | Agric. indus. country | + + | – | + | Colombia, Iran, Ecuador |
| Argentina, Uruguay | + | + + | – | Indus. agric. country | + + | + | – | Chile, Finland |
| USA, Australia | + | + + | – – | Indus. country | + | + + | – – | Great Britain, FRG |
| | *B. Quantitative Combination* | | | | | | | |
| see above | + + | + | – – | Agric. country | – – | + + | – – | see above |
| | + + | + | – | Agric. indus. country | – – | + | – | |
| | + | – | + | Indus. agric. country | – | – | + | |
| | – | – – | + + | Indus. country | – | – – | + + | |

Costs and quantity-input of production factors are – – very low, – low, + high, + + very high.

the relative prices of land rent, wages and interest and alone ensures the cheapest production.

In a farm organized on commercial principles the money value of the marginal product of each variable production factor should equal its costs. As the marginal productivity falls if this production factor is gradually increased in relation to the other two, then, other things being equal, each production factor must be used more sparingly as its price becomes higher. On the other hand, the cheapest production factor should be used in the largest quantity. Progressive economic development, which is generally achieved through increasing industrialization, has the result that capital goods gradually become cheaper, while labour becomes more and more expensive. As a result, human labour is increasingly equipped with capital goods. The choice of capital goods in the course of economic development depends on the scarcity of land. In thinly populated agricultural countries, the primary aim will be to replace manual labour. Initially, this is much more important than economizing in land. Under these circumstances labour saving capital goods are of primary importance, and the mechanization phase precedes the intensification phase. Only at a later stage, when land also becomes scarce, is it also necessary to use yield increasing farm inputs.

The situation in overpopulated agricultural countries is quite different. Here, land becomes scarce sooner than labour, so that priority in expenditure on capital goods for agriculture must be given to yield increasing farm inputs which economize on the use of land. In this case, intensification precedes mechanization, which only occurs later in the development process when farm labour has also become scarcer and dearer, with an advanced stage of industrialization.

In short it is clear that the minimum cost combination in agriculture varies very widely and that it must undergo great changes in the course of overall economic development. The assessments of the trend of change in minimum cost combination of a developing country will vary according to whether it starts off as an over-populated agricultural country or a thinly populated agricultural country. As a result the importance of technical progress must also vary for developing countries of different structures. Innovations in mechanization will primarily benefit thinly populated agricultural countries, whereas biological innovations chiefly ease the future development of overpopulated agricultural countries.

## Development of Farming Systems in the Tropical Climatic Zones

In the course of overall economic development, land becomes scarcer and therefore has to be farmed more and more productively, i.e. intensively. A country in the transition from being a thinly populated agricultural country to a densely populated one initially has a growing number of workers available although all forms of capital goods remain very expensive. In this phase of development the necessary intensification of agriculture must take place through increased use of labour. When industrialization reaches an appreciable level labour as well as land becomes scarcer and more expensive while industrially produced capital goods become cheaper. In this phase, further intensification must take place through greater use of capital.

In the course of overall economic development, land productivity must be increased first and at a later stage, labour productivity. The trio of intensity stages that satisfies this requirement is as follows:

extensive ————▷ labour intensive ————▷ capital intensive.

**Rain forest climate**

In the permanently wet tropical rainforest climate, the farming systems will probably always remain decidedly plant producing systems without livestock, and change in their quantitative relationship in the course of economic development as follows:

1st stage:    Wild shifting cultivation
2nd stage:    Controlled shifting cultivation with natural fallow forest
3rd stage:    Controlled shifting cultivation with artificial fallow forest
4th stage:    Controlled shifting cultivation with plough-cultivation
5th stage:    Crop rotations including pulse crops fallow
6th stage:    Crop rotations including grass fallow
7th stage:    Ley farming
8th stage:    Perennial crops.

This sequence of stages in the development of farming systems satisfies the requirement of first increasing land productivity and later also increasing labour productivity. It is, therefore, the rule although it does not preclude the possibility of some stages being omitted when the economic development is very rapid and dynamic. The sequence of stages can also change to some extent. In parts of Nigeria, oil palm cultivation directly replaced the wild shifting cultivation omitting intermediate stages. The same thing happened with cocoa production in the south of Ghana. The favourable geographical location of both these countries for ocean borne trade, made it possible to take advantage of high world market prices at that time. The combination of artificial forest fallow with plough cultivation has its problems and this stage is passed over in many countries. Foreign trade often makes it possible to start growing perennial crops immediately, even when the domestic market has not yet developed sufficiently to absorb the animal products from the intermediate stage of ley farming.

In general, however, the sequence of stages of farming systems set out above must occur in the course of economic development. It is a realistic sequence, because it is based on our theory that the increase in the proportions of land used, with accompanying increases in yields per ha must initially be achieved by the use of more labour and later on by using more capital. The changeover from wild to controlled shifting cultivation initially only increases the proportion of land which is cultivated, i.e. it only increases the amount of labour used. Although the changeover from natural to artificial afforestation of fallow requires capital in the form of planting materials, this is of the kind that is produced within the enterprise by labour. Capital expenditure only increases considerably at the stage when hoe-cultivation gives way to plough cultivation and capital is needed for traction power and implements. When forest fallow is replaced by pulse fallow, seed capital is also required, and with the changeover from pulse crops fallow to grass fallow, fertilizer capital is required as well. At the latest stage both the highest forms of land use, ley farming and perennial crops, require considerably more capital in the form of livestock or permanent plant-stock.

**Humid savannas**

In the humid savanna, the importance of farming systems will probably change as follows in the course of economic development:

1st stage:    Shifting cultivation in the forest
2nd stage:    Shifting cultivation in the bush
3rd stage:    Crop rotations including grass fallow
4th stage:    Rainfed agriculture without fallow and without livestock
5th stage:    Rainfed agriculture without fallow but with livestock
6th stage:    Irrigated agriculture without fallow but with livestock

Here development moves initially towards an increase in the proportion of land used, with a continuing reduction in the number of fallow years and a consequent change in the fallow vegetation from forest to bush and then to grass. Later on, the pressure to increase land productivity leads to an attempt being made to do entirely without fallow. It is soon found that permanent rain fed agriculture requires livestock to provide manure and livestock husbandry is now made economic because general development has resulted in a higher demand for animal products.

Finally livestock husbandry with its requirement for a seasonal adjustment of feed supplies is itself a contributory factor towards the last decisive step to irrigated agriculture. The humid savanna is ideal for this, because there are abundant ground-water supplies available in relation to the relatively low demand for water, with seven to nine humid months in the year. In addition rivers carrying water all through the year make dam-water supply possible and economical. So long as land productivity is the primary concern and labour is still cheap, a variety of irrigation systems will be adopted. From a certain wage level upwards, overhead spray irrigation will be more economic, because it requires much less labour and in the course of development fixed capital investment becomes cheaper. With given stages of development and capital costs, the wage level in thinly populated countries is higher than in densely populated countries. Thus the rates of replacement of labour by capital may ensure that for countries with a sparse population overhead irrigation will replace irrigation by the basin or check system even at the agricultural-industrial stage of development while for those with a dense population the change will not occur until they reach the industrial-agricultural stage. At that stage overhead irrigation in combination with fertilizers, plant protection etc., provides high labour productivity as well as high land productivity.

**Dry savannas**

Although the dry savannas have a greater need for irrigation the lack of water will make irrigated agriculture less acceptable. The most important stages of change in types of farming in the course of economic development are as follows:

1st stage:   Extensive grazing systems
2nd stage:   Savanna shifting cultivation
3rd stage:   Dry farming systems alongside grazing systems
4th stage:   Development of crop rotations through combining the growing of fodder and pulse crops with extensive grazing systems
5th stage:   Advance of grazing systems over rain-fed agriculture.

In Stage 4 fodder growing is initially undertaken chiefly for purposes to do with soil fertility and not so much for the feeding of stock. At first ley farming is introduced primarily to support the growing of cereals, mostly millet and maize, rather than to help livestock enterprises during periods of feed shortage. This does not of course preclude the growing of field fodder to provide grazing during droughts or for the production of hay. Such field fodder crops provide the connecting link between the livestock branches (cattle fattening) and the arable growing of maize, millet or barley. The growing of field fodder makes it possible to establish crop rotation in cereal growing and to widen the fodder base for cattle in the drought period. The system provides humus in the form of crop residues and roots for arable farming and winter feed for grazing cattle, and the cattle convert part of this feed into manure, which in turn benefits the grain crops. In this way the two branches of farming for the first time form an association, an integrated whole, an operational system from the production of cattle for slaughter and cereals growing which had previously existed side by side but in isolation. In the course of further development, more productive field crops such as peas, groundnuts, sunflowers, sesame and possibly even cotton, etc., are included in the farming system, which thus advances further and further over the natural pastureland and ultimately takes over the entire area where natural conditions of slope of terrain, depth of topsoil, groundwater conditions, etc., are suitable for arable farming.

**Thornbush steppe (shrub)**

The thornbush steppe lies outside the zone which is capable of rain-fed farming and development can only take place within extensive grazing systems. In accordance with the guideline

extensive ————▷ labour intensive ————▷ capital intensive

the change in the form of farming in the course of overall economic development is as follows:

1st stage:   Nomadic shepherds
2nd stage:   Stationary grazing with seasonal migration to other pastures (transhumance)

3rd stage:    Bridging the drought period from the farm's own fodder reserves on
              pastures and by loss of deposited fat
4th stage:    Increase in the number of watering places
5th stage:    Increase in the subdivisions of pastures
6th stage:    Buying in fodder
7th stage:    Production of own fodder for the drought period

This sequence of stages in the use of capital is induced not only by an improvement in
the exchange value of animal products against farm working capital on the market,
but also by an improvement of infrastructure which reduces the economic distance of
farms from the market. The improved feed balance also makes it possible to change
the forms of livestock husbandry in a way appropriate to the natural conditions of
production so as to increase productivity.

## Radical Changes in the Future Agricultural Landscape of the Tropics

Finally, if we take a glimpse into the more distant future for these four main
categories of tropical farming competitive changes and shifts in the location of the
farming systems in the equatorial belt are likely to produce the following picture:

The zones which will expand include (1) farming with perennial crops, because
world markets for most of their products are expanding. They are intensive branches
of farming and the inner tropics urgently need them to preserve their endangered soil
fertility; (2) irrigated farming, because national economies are increasingly developing their water supplies by large-scale dam projects etc., because irrigation increases
the productivity of both land and labour and because the basin or check system of
irrigation preserves soil fertility in a similar way to the original vegetation.

The zones which will contract include (1) rainfed farming. In the rainforest belt it will
be replaced by farming with perennial crops; in the humid savannas by irrigated
agriculture and near the agronomic drought boundaries by the advance of grazing
systems. (2) Ranching, will, on balance contract. Although it will gain, land in the
marginal zones of modern rain-fed farming which are too low yielding to be worth
cultivating with rising wage levels, this gain will be more than offset by loss of acreage
in semi-deserts and thornbush steppe, from which extensive grazing systems will
probably have to withdraw because productivity is too low.

The arable farming zone will thus be reduced from two sides—from the margins of
the tropics by grazing systems and from the equator by the cultivation of perennial
crops. Thus, larger and larger portions of the tropics will be able to revert to some-
thing nearer their natural vegetation cover. The tropical margins will return to
grassland, and the equatorial belt to tree and bush cover, even though this will be a
cultivated landscape which has developed from the original natural landscape.

## Futurological Aspects of Agricultural Technology

In conclusion I would like to touch on various futurological aspects of the
agricultural industry, since progress in the provision and efficient utilization of water
resources is only one way of extending the food base and must be viewed against its
overall background.

The forces which bring about agricultural development can be defined simply as technical advances, or price and cost changes. Technical developments, in so far as they are actually innovations, are usually unforeseen and sudden. Since it is impossible to predict such completely new technical developments even approximately, we will consider some developments made possible by existing technology which could eventually offer prospects of yield or economy increases. These examples are taken from Thiede's 1972 book on the revolution in agrarian technology.

**Farm machinery**

The Kuratorium für Technik und Bauwesen in der Landwirtschaft e. V. (German Board of Trustees for Agricultural Technology and Construction Systems) expects that machines with tractive powers of 100–200 h.p. and operating widths of five to six metres will become of major importance. Self-propelled specialized machines, (like combine harvesters) will in future be used for ploughing, sowing, fertilizer spreading and other treatment and harvesting operations. Mobile drying plants for green fodder are already in use. A Hohenheim study suggests that by 1990 technical advances will make it possible to reduce man hours per ha for cereal farming to only six and for root crops to only 30. Operating velocities and widths of field-cultivation implements would increase by about 15% per decade, and would be accompanied by an increase of 40–50% in the average power of harvesting machines for cereals and root crops.

By the year 2000 automatically controlled work implements could be available. By that time only the ears of cereals would be harvested. Sugarbeet juice would be extracted in the beet field. Experts writing in the GFR Agra-Europe in 1973 even suggest that solar and nuclear energy will be used to drive farm machinery.

Only 25 years ago one farmer could feed eight to ten persons, nowadays he can feed 26; it will not be very long now before the European farmer will be able to feed 50.

**Crop production**

All sorts of radical technical changes are being made in actual crops. Attempts are now being made to keep fruit stalks so small that labour can be saved by "mowing" the fresh branches growing out of the ground each year instead of picking the fruit. The expected hybrid grades of sorghum millet could produce significant changes in the dry summer Mediterranean area, while parts of the tropics and sub-tropics could be similarly affected by the high yielding varieties of wheat, maize and rice. Dr. Borlaug was awarded the 1970 Nobel Peace Prize in honour of his work in this field. In the case of bacteria and fungi, molecular biology has succeeded in creating completely new organisms through gene adaptation. Similar success in the higher vegetable species would amount to a revolution in agrarian technology.

Futurologists do not discount the possibility of some day utilizing the vast energy of the interior of the earth. The great hot water ocean under the west Siberian plain would then enable this region to serve the world as a major new agrarian producer.

**Livestock production**

In poultry farming the difficulties involving smell nuisance and excrement disposal in factory farming, have resulted in discussion of a "throw-away stall" which is only used for one laying period and then acts as a roofed manure heap. In Japan farmers can already call herds of cows back to their cowshed by radio, through installing a radio receiver in the horn points of the leading animal. In pig breeding, trials are in progress with motherless rearing, with automatic piglet feeding from birth, in order to produce virtually three litters per year. Animal breeding trials are employing operative transplantation of fertilized egg cells from the most valuable mother animals, into genetically less valuable foster mothers. In this way it is hoped to distribute high quality female genotypes more rapidly, with accompanying improvement of the male genotype through artificial insemination. There are also prospects of success from experiments to separate X- and Y- sperms in artificial insemination and thus control the sex of the offspring.

In cattle feeding, synthetic urea fodder for cattle is being introduced. This makes it possible to use nitrogen from the air to replace vegetable proteins. Already the USA uses 600 000 t of urea fodder per year, and this corresponds to a saving of about 2 million ha of soya beans. "Fodder from petroleum" is already a reality. Single-cell protein feeds in the form of yeasts and bacteria are produced from petroleum (or natural gas or hydro-carbons) through biosynthesis. One day these aids will overcome the protein shortage still troubling the fodder economy of many farms. Finally mention must be made of the efforts being made to increase production of non-agricultural foodstuffs, not only to tap the food reserves of the seas but also to develop synthetic foodstuffs.

## More Food Production from the Seawater

Like Plato and Aristotle, Montesquieu and Mirabeau, Franklin and Malthus before them, the United Nations have called for action against world hunger. Such action will only be effective if water resources utilized in food production are exploited less and are managed more effectively than hitherto. Management, however, means husbandry on the basis of deficient resources and the creation of reserve resources. It is hardly possible to realize nowadays that centuries ago land was a free commodity, and we are still equally incapable of imagining that air could be anything other than a free commodity. And yet in many parts of the world even sea water is ceasing to be a free commodity and is becoming a scarce commodity of economic importance.

**Cultivation of the seas**    The intensive utilization of seawater like land for purposes of cultivation, has so far only been carried out in isolated cases such as oyster breeding on the French Mediterranean coast. In the USA oyster larvae are fed with the products of accompanying algae cultures. In Japan oyster breeding is undertaken on a grand scale. Meske (1973) points out that in Australia aquatic cultures yield US$2.7 million. The fish farms in the Javan shallow coastal waters, and similar fish farms breeding oysters in Japan are described by Thiede (1972). All these are, however, only pinpricks in the exploitation of the oceans, bearing in mind that the sea covers some 70% of the earth's surface and is the habitat of about 80% of its animal life.

We are still very far from exploiting the seas and are still only at the stage of appropriating what nature provides without our help. As regards the oceans we are still at the hunting and gathering stage of development by which Neanderthal man in the early stone age utilized the land masses of the earth, a way of life still followed by bushmen, American Indians and eskimos even at the present day. Modern man has not progressed beyond this earliest appropriating stage so far as the exploitation of the sea is concerned, despite having lived through the new stone age, the development of agriculture and the urban and the industrial revolutions.

**Sea farming**

There is still no consensus of opinion as to how the seas could be rationally exploited in future for purposes of food production. Futurologists have put forward the idea that enormous volumes of krill (a higher plankton species found in the Antarctic Ocean) could be applied in protein production, using large sieve tanks driven by nuclear energy. Thiede (1972) considers it quite feasible that krill could become a significant contributor to human food supply in the next decade or so.

Other ideas about future sea farming are less optimistic. Projects have been devised for growing fodder through suitable planting of seabeds, new techniques for breeding fish, and artificial protection of young fish against their enemies. In principle, all these projects are faced with more extreme forms of the same kind of difficulty as faces intensive agrarian activities at the dry limits for stock raising. Keeping animals on a lead involves an extremely high expenditure. Fish shoals can be fenced in by air bubbles, sound waves or electric fields. The station on the Isle of Man which breeds about one million fish a year is an example of this. The Japanese have developed techniques for rearing numerous species of sea fish in floating wire cages in the bays. This permits effective feeding and harvesting (Meske 1973).

**The desalination of sea water**

The exploitation of sea water is possibly one of the main keys to overcoming the world food problem. It may perhaps one day prove possible to breed crops which will tolerate a greater salt concentration than those we have now. In North Africa and the Near East the most salt compatible crop is the date palm, to such an extent that around Basra the date groves can be irrigated by water from the Persian Gulf. Although these waters are in fact from the Tigris and the Euphrates, they are relatively salty in the estuary. Crops are also irrigated by such brackish waters on the Gulf Coast of Iran near Abadan. Desalination of sea water is however a more promising solution. It could open up an enormous reservoir for irrigation to supplement the water resources we now have in impounded river water and ground water. Over the world as a whole, more than 1 000 000 m$^3$ of fresh water is produced from sea water every day and it is estimated that this figure could be increased tenfold over the next 10 years. (Fishbek 1971). The surplus heat from atomic power stations can provide an economic source of heat for this process.

The first European seawater desalination plant for public drinking water supply was brought into use in Heligoland in 1972. It yields 800 m$^3$ of absolutely fresh water per day, which, mixed with 300 m$^3$ of brackish water, supplies sufficient drinking water for the island even at periods of peak demand. The cost/benefit ratio is very favourable, since previously drinking water had to be brought from the mainland by tanker at a cost of DM27 per/m$^3$. Seawater desalination has reduced the cost of drinking water to DM2 per/m$^3$ (Schwenke 1972).

If large installations of this type were some day to be installed on the sea coasts of underdeveloped countries with dry climates, such as North Africa or Arabia for irrigation purposes the resulting changes in the location of world agricultural production, the structure of world agricultural trade and the growth rate of the national economies affected could be infinite.

**Control of rainfall**

Technically there are many possibilities of weather control. Clouds can be made to give rain or disperse. Hail, storms and cyclones can be warded off or deflected, fog can be cleared (Vester, 1968). The possible significance of such techniques in coping with dry periods, ensuring satisfactory harvesting weather, etc. can be judged from an example.

In the dry climate of South West Africa the winds blow from the east, as a result of its geographic location in the southern trade wind region. The moisture brought by these winds from the warm Indian Ocean is largely lost as they rise over the eastern edge of the Drakensberg Mountains. They thus pass over the Kalahari as very dry winds and only supply a small volume of water to South West Africa.

If the clouds in the north east part of the country were made to drop their rain artificially, this would amount only to a redistribution of the rain. The farmers in the north east would benefit at the expense of those in the south west. The agronomic drought limit in the north east (about 500 mm of rain) and the Namibia Desert in the south west would be enlarged relative to the natural pasture zone. This could be in the national economic interest, as it would concentrate agriculture in smaller but more productive zones, but the civil law and social consequences would be hard to imagine.

If it were to prove possible to eliminate rainfall, before the coast is reached, from the west winds coming from the Atlantic which release their rain over the cold Benguella current, the situation would be different. This would not merely redistribute, but would increase the rainfall in South West Africa and would be beneficial throughout the country. The nonproductive rainfall over the ocean would be replaced by highly productive rainfall over land masses suffering from drought.

Israel is experimenting with the diversion of rainwater in quite a different way. There they are spraying slopes with plastic films in order to increase the volume of rainwater retained. A 30–50 ha catchment area is required for every ha of farmland (cereals, vegetables, peaches) (Voegelin 1968).

# Bibliography

*Agra Europe* (1973) No.25

Andreae, B. (1965) [Land fertility in the tropics. Utilization and maintenance. Farm management considerations for work in developing countries.] Die Bodenfruchtbarkeit in den Tropen. Nutzbarmachung und Erhaltung. Betriebswirtschaftliche Überlegungen für die Arbeit in Entwicklungsländern. Hamburg; Berlin, German Federal Republic; Paul Parey 124pp.

Andreae, B. (1966) [Pasturing in southern Africa. Studies of Location and development theory of the agricultural geography of the tropics and sub-tropics.] Weidewirtschaft im südlichen Afrika. Standorts- und evolutions-theoretische Studien zur Agrargeographie der Tropen und Subtropen. *Geographische Zeitschrift, Beihefte Erdkundliches Wissen* No.15, 50pp.

Andreae, B. (1974) [The diversification and specialization of farming in the tropics.] Diversifizierung und Spezialisierung der Farmwirtschaft im Tropenraum. *Berichte über Landwirtschaft* 52, 497–511.

Andreae, B. (1974) [Which farm will survive? Preserving the farm business by developing it.] Welcher Hof wird überleben? Betriebserhaltung durch Betriebsentwicklung. Hamburg; Berlin, German Federal Republic; Paul Parey 194pp.

Andreae, B. (1977) [Agricultural geography.] Agrargeographie. Berlin; New York, USA; Walter de Gruyter 332pp. (English language edition forthcoming, New York 1980).

Andreae, B. (1978) [Agricultural regions under local stress.] Agrarregionen unter Standortstress. Kiel, German Federal Republic; Verlag Ferdinand Hirt 78pp. (*Geocolleg* No.6).

Andreae, B. (1978) The minimum cost combination in agriculture. *GeoJournal* 2 (3), 203–214.

Andreae, B.; Greiser, E. (1978) [Structures of the German agricultural landscape.] Strukturen deutscher Agrarlandschaft. *Forschungen zur deutschen Landeskunde* No. 199, 124pp. (Edition 2).

Angladette, A.; Deschamps, L. (1974) [Problems and perspectives of agriculture in tropical countries.] Problèmes et perspectives de l'agriculture dans les pays tropicaux. Paris; Maisonneuve et Larosse 770pp.

Bader, F.J.W. (1965) [Uganda. A model of African possibilities.] Uganda. Ein Modell afrikanischer Möglichkeiten. *Geographische Rundschau* 17, 83–96.

Barber, W.J. (1961) The economy of British Central Africa. A case study of economic development in a dualistic society. London, UK; Oxford University Press 271pp.

Besters, H. (1955) [Theories of economic development.] Theorien der wirtschaftlichen Entwicklung. In: Entwicklungpolitik. Handbuch und Lexikon. (Edited by H. Besters and E.E. Boesch). Stuttgart, German Federal Republic; Kreuz-Verlag 243–304.

Blanckenburg, P. von (1965) [African peasant farming on the way to a modern agriculture.] Afrikanische Bauernwirtschaften auf dem Weg in eine moderne Landwirtschaft. *Zeitschrift für ausländische Landwirtschaft, Sonderheft* No.3, 112pp.

Bloodworth, D. (1970) An eye for the dragon. South-east Asia observed, 1954–1970. London, UK.

Brown, L.H. (1968) Agricultural change in Kenya: 1945–1960. *Stanford University Food Research Institute Studies* 8, 33–90.

Bruton, H.J. (1965) Principles of development economics. Englewood Cliffs, USA; Prentice-Hall 376pp.

Clayton, E.S. (1964) Agrarian development in peasant economies: some lessons from Kenya. Oxford, UK; Pergamon Press 154pp.

Eckaus, R.S. (1960) The factor-proportions problem in underdeveloped countries. In: Agarwala, A.N. and Singh, S.P. (*Eds*) The economics of underdevelopment. New York, USA; Oxford University Press 348–378.

Fischbeck, K. (1971) [Fresh water from the sea.] Süsswasser aus dem Meer. In: Bild der Wissenschaft. Stuttgart, German Federal Republic.

Fuchs, H. (1973) [Systems theory and organization.] Systemtheorie und Organisation. Wiesbaden, German Federal Republic; Gabler.

Galenson, W.; Leibenstein, H. (1955) Investment criteria, productivity and economic development. *The Quarterly Journal of Economics* 69, 343–370.

Hagen, E.E. (1968) The economics of development. Homewood, Illinois, USA; Nobleton, Ontario, Canada; Irvin 536pp.

Han Suyin (1973) The morning deluge. Mao Tsetung and the Chinese Revolution. London, UK.

Healey, D.T. (1964) Agricultural economics in some African countries. *International Journal of Agrarian Affairs* 4 (4) 250–286.

Heidhues, T. (1976) [Agricultural policy 1: prices and incomes policy.] Agrarpolitik 1: Preis- und Einkommenspolitik. In: Handwörterbuch der Wirtschaftswissenschaften, Vol.I. 107–128. Göttingen, German Federal Republic; Gustav Fischer.

Herlemann, H.H. (1954) [Stages in the technical development of agriculture.] Technisierungsstufen der Landwirtschaft. *Berichte über Landwirtschaft* 32, 335–342.

Herlemann, H.H. (1961) [Principles of agricultural policy.] Grundlagen der Agrarpolitik. In: Kade, G. (*Ed*) Vahlens Handbücher der Wirtschafts- und Sozialwissenschaften. Berlin; Frankfurt/Main, German Federal Republic; Franz Vahlen GmbH, 191pp.

Hohnholz, J.H. (1978) The rural area's role in the development of Southeast Asia with special reference to Thailand. *Materialien zum Internationalen Kulturaustausch* 6, p.154 et seq.

Humlum, J. (1977) [China masters hunger.] China meistert den Hunger. Kiel, German Federal Republic; Ferdinand Hirt 136pp. (Geocolleg 4).

Johnston, B.F. (1964) The choice of measures for increasing agricultural productivity: a survey of possibilities in East Africa. *Tropical Agriculture* 41, 91–113.

Johnson, E.A.J. (1970) The organization of space in developing countries. Cambridge, Massachusetts, USA; Harvard University Press 452pp.

Junghans, K.H. (1968) [Influence of industrialization on the geographical mobility and cultural adjustment of traditional agricultural societies in South Asia.] Einfluss der Industrialisierung auf die geographische und geistige Mobilität traditioneller Agrargesellschaften in Südasien. *Geographische Rundschau* 20 (11) 424–431.

Junghans, K.H.; Nieländer, W. (1971) [Indian peasants on the way to market. The example of Rourkela.] Indische Bauern auf dem Wege zum Markt. Das Beispiel Rourkela. *Wissenschaftliche Schriftenreihe des BMZ* 20, 187pp.

Kindleberger, C.P. (1965) Economic development. New York, USA; London, UK; Sydney, Australia; Toronto, Canada; McGraw-Hill 425pp.

Kolb, A. (1977) [Understanding the contemporary history of China.] Zum zeitgeschichtlichen Verständnis Chinas. *Geographische Rundschau*, 29, 39–43.

König, H. (1970) Approaches and problems of growth theory. *Economics* 1, 70–87.

Krelle, W. (1964) [Investment and growth.] Investition und Wachstum. *Jahrbuch für Nationalökonomie und Statistik* 176, (5) 1–22.

Kruse-Rodenacker, A. (n.d.) [Basic questions of development planning.] Grundfragen der Entwicklungplanung. *Schriften der deutschen Stiftung für Entwicklungsländer* 1, 298pp.

Kruse-Rodenacker, A. (1965) [Selection criteria in choosing industrial projects in developing countries.] Selektionskriterien zur Auswahl von Industrieprojekten in Entwicklungländern. In: Die Stellung der Landwirtschaft und Industrie im Wachstumsprozess der Entwicklungländer. *Schriften des Vereins für Socialpolitik* 43, 57–69.

MacEwan, A. (1971) Development alternatives in Pakistan. Cambridge, Massachusetts, USA; Harvard University Press 211pp.

Manshard, W. (1968) [Agricultural geography in the tropics. An introduction.] Agrargeographie der Tropen. Ein Einführung. (B.J.-Hochschultaschenbücher 356/356a) Mannheim, German Federal Republic; Zürich, Switzerland; Bibliographisches Institut AG 307pp.

Masefield, G.B. (1962) A handbook of tropical agriculture. Oxford, UK; Oxford University Press 195pp.

Massow, H. von (1964) [Opportunities for industrialization in Tanganyika.] Industrialisierungsmöglichkeiten in Tanganyika. Hamburg, German Federal Republic; Afrika-Verein, Technisch-Wirtschaftlicher Dienst 196pp.

Meske, C. (1973) [Aquaculture of economically useful warm water fish.] Aquakultur von Warmwasser-Nutzfischen. Stuttgart, German Federal Republic; Eugen Ulmer 163pp.

Ochse, J.E. et al. (1966) Tropical and subtropical agriculture, Vols.I and II reprint. New York, USA; London, UK; Macmillan 1446pp.

Predöhl, A. (1970) The world economy and transport. *Economics* 1, 55–69.

Preiser, E. (1971) Economic growth as a fetish and a necessity. *Economics* 3, 128–140.

Schneider, E. (1971) Economic growth and economic order. *Economics* 3, 101–108

Schwenke, C. (1972) [Fresh water from salt water.] Aus Salzwasser wird Süsswasser. *VDI-Nachrichten* 26, p2 et seq.

Teo Chris, K.H.; Atanasiu, N. (1975) Increased land productivity or expanded area as a means for increasing food production in developing countries. Workshop on Energy, Resources and the Environment. Penang, Malaysia.

Thiede, G. (1972) [The revolution in agricultural technology and agriculture in the future.] Agrartechnologische Revolution und zukünftige Landwirtschaft In: Die Zukunft des ländlichen Raumes, Part II. *Forschungs- und Sitzungsberichte der Akademie für Raumforschung und Landesplanung*, 83, 1–24.

Thiede, G. (1973) [The agricultural sector between reality and futurology.] Agrarwirtschaft zwischen Realität und Futurologie. Lecture, Düsseldorf, German Federal Republic, 3rd December, 1973. Hamburg, German Federal Republic; Alfred C. Toepfer 23pp.

USA, Department of Agriculture (1969) Economic progress of agriculture in developing nations 1950–68. *Foreign Agricultural Economic Report, Economic Research Service* No.59, 180pp.

USA, Department of Agriculture (1971) The agricultural situation in the Western hemisphere. Review of 1970 and outlook for 1971. *Economic Research Service-Foreign* No.312, 27pp.

Vester, F. (1968) [Building blocks of the future.] Bausteine der Zukunft. Frankfurt am Main, German Federal Republic.

Voegelin, D. (1968) [The future of the water industry has already begun in Israel.] Die Zukunft der Wasserwirtschaft hat in Israel schon begonnen. *Mitteilungen der DLG* 83, 440–443.

# Subject Index

# Geographical Index